GENETICS - RESEARCH AND ISSUES

GENETICS RESEARCH SUMMARIES

GENETICS - RESEARCH AND ISSUES

Additional books in this series can be found on Nova's website under the Series tab.

Additional E-books in this series can be found on Nova's website under the E-book tab.

GENETICS - RESEARCH AND ISSUES

GENETICS RESEARCH SUMMARIES

LYNDSAY E. HEWITT
AND
JOHN W. O'NEIL
EDITORS

NOVA BIOMEDICAL

New York

Library of Congress Cataloging-in-Publication Data

ISBN 978-1-62257-646-3

Published by Nova Science Publishers, Inc. † New York

Contents

Preface

This new book is a continuation of Genetics Researcher Biographical Sketches and Research Summaries, which compiles research summaries from a number of different focuses in this important field.

In: Genetics Research Summaries
Editors: L. E. Hewitt and J. W. O'Neil
ISBN: 978-1-62257-646-3

Chapter 1

Domain-Oriented Analysis for Understanding Molecular Interactions and Translating Animal Genetic Models into Neuropsychiatric Disorders: Focus on Serotonin Transporter and BDNF*

Allan V. Kalueff*[#] *Renee F. Ren-Patterson, Justin L. LaPorte and Dennis L. Murphy
Laboratory of Clinical Science,
National Institute of Mental Health, NIH, Bethesda, MD, US

Research Summary

Animal experimental models of brain disorders represent a valuable tool in today's biological psychiatry. Various genetic animal models, based on selectively bred, hybrid, gene-targeted or transgenic animals, are widely used in neuroscience research for screening neuroactive drugs, testing neurobiological hypotheses and finding candidate genes for stress-related brain disorders. Numerous data indicate that brain disorders in many cases represent overlapping pathogenetic pathways with common genetic determinants and clinical manifestations. Together, this raises the possibility that a *combination* of several distinct but

* A version of this chapter was also published in Behavioral Models in Stress Research, edited by Allan V. Kaleuff and Justin L. LaPorte published by Nova Science Publishers, Inc. It was submitted for appropriate modifications in an effort to encourage wider dissemination of research.

Corresponding author: E-mail address: avkalueff@inbox.ru. Phone: 1-301-594-0393. Laboratory of Clinical Science, Building 10, Room 3D41, National Institute of Mental Health, Bethesda, Maryland 20892-1264, USA.

interacting domains, and their *interplay*, may contribute to clinical and experimental phenotypes. This also implies that a closer in-depth analysis of different domains may be needed for developing new, clinically relevant genetic experimental modeling of neuropsychiatric disorders.

In this chapter, we will discuss how analysis of different behavioral and physiological domains in genetic animal models may optimize further experimental research in this field. We will also argue that these approaches can be used to pursue another far-reaching goal: to elucidate complex brain processes and their potential molecular underpinnings. To illustrate the utility of this approach, we will specifically focus on two key brain molecules – serotonin transporter (SERT) and brain-derived neurotrophic factor (BDNF), both implicated in multiple neuropsychiatric disorders.

In: Genetics Research Summaries
Editors: L. E. Hewitt and J. W. O'Neil
ISBN: 978-1-62257-646-3

Chapter 2

Understanding the Stress Response through Mouse Genetics*

Amy F. Eisener-Dorman and Valerie J. Bolivar
Wadsworth Center, David Axelrod Institute, Albany, NY, US
School of Public Health, Department of Biomedical Sciences, State University of New York at Albany, Albany, NY, US

Research Summary

Stress helps organisms adapt to the challenges of the environment, however, too much of it can be harmful. High levels of stress have been linked to a number of physiological and psychological conditions including, but not limited to, psychosomatic, immunological, cardiovascular, and affective disorders. It is well known that the specific type of stress and stressor characteristics (e.g., controllability, predictability, ambiguity, chronicity, and intermittence) influence the stress response. While some individuals appear to be susceptible to stress-related disorders, others are not, and there seems to be much variability in response to relatively equivalent stressors. Why do some individuals have stronger biological responses to stress than others? Research indicates that a combination of experiential factors, resource characteristics, and organismic variables (e.g., species, strain, age, and sex) influence the magnitude of the stress response.

Through rodent research, we have gathered a great deal of information about the effects of variables such as age, sex, and strain on the nature and impact of stressors. Although research with rats allows exploration of the neurobiological consequences of stress, as well as the effects of age and sex on these processes, mouse research has allowed for more advanced genetic approaches and a more thorough examination of the role of genetic factors on the stress response. With the readily available transgenic/targeted gene technologies and the recent

* A version of this chapter was also published in Experimental Animal Models in Neurobehavioral Research, edited by Allan V. Kaleuff and Justin L. LaPorte published by Nova Science Publishers, Inc. It was submitted for appropriate modifications in an effort to encourage wider dissemination of research.

sequencing of the mouse genome, the mouse has become the model of choice for many behavior geneticists and neuroscientists in the study of human disorders and behaviors, including stress. Research involving inbred mouse strains, in addition to transgenic and gene knockout mouse models, has helped elucidate the role of specific genes on the stress response and determine the impact of stress on other pathways. Mouse model research has also helped us understand the gene x environment interactions that influence stress responses.

In this chapter, we will examine the impact of mouse genetics in a number of areas of stress research, as well as discussing the important issue of genetic background and the role it may play in any stress-related phenotype observed in a mutant mouse. A number of genes investigated in mouse model systems have helped shape our understanding of the stress response and illustrate how stress affects a diversity of biochemical and molecular pathways. Some of the genes we will review have direct relationships to the stress response (e.g., *Crhr1*, *Crhr2*, *Avpr1a*, *Avpr1b*), whereas others are more indirectly related and the pathways involved have not yet been well characterized. Nevertheless, mouse genetics has helped us to understand the plethora of pathways related – directly or indirectly – to the stress response. It is evident that the effects of stress are widespread in the organism and that much remains unknown of the intricacies involved. Due to the magnitude of pathways involved, in our review we will highlight only a few of them. We will begin our investigations with the most obvious genes involved in stress – those genes intimately involved in the hypothalamic-pituitary-adrenal (HPA) axis.

In: Genetics Research Summaries
Editors: L. E. Hewitt and J. W. O'Neil

ISBN: 978-1-62257-646-3

Chapter 3

Apolipoproteins in Primary Antiphospholipid Syndrome

Mirjana Bećarević[1,2] ***and Nada Majkić-Singh***[1,3]
[1]Institute for Medical Biochemistry, Clinical Center of Serbia, Belgrade, Serbia;
[2]Medical Faculty, Novi Sad, Serbia
[3]Pharmaceutical Faculty, Belgrade, Serbia

Research Summary

The antiphospholipid syndrome (APS) is an acquired thrombotic disorder characterized by venous and arterial thromboses and/or spontaneous abortions and repeated detection of antiphospholipid antibodies. The antiphospholipid syndrome may be associated with another autoimmune disease (secondary APS = SAPS), or unrelated to an underlying disease (primary APS = PAPS).

Patients with antiphospholipid syndrome have increased risk for atherosclerosis and cardiovascular complications. Lipids through their effects on the coagulation and fibrinolytic system may contribute to the development of venous thromboses, and this association has been investigated in a few studies with conflicting results. Thus, clinical research has focused on the development of novel biomarkers that might detect patients prone to thrombosis, which is the main complication of atherosclerosis, and some of them are apolipoproteins.

Decreased apolipoprotein AI (apo AI) levels, and increased apolipoprotein B (apo B) levels are risk factors for development of atherosclerosis. Apolipoprotein (a) (apo (a)) is a glycoprotein linked by single disulphide bridge to apo B in the lipoprotein (a) (Lp(a)) particle. Despite the fact that elevated Lp(a) concentrations have been reported in autoimmune diseases, significance of apolipoproteins concentrations has not been widely analyzed in patients with PAPS.

Elevated apo B levels were present in PAPS patients with history of thromboses of peripherial arterial blood vessels in comparisons to the patients without abovementioned clinical finding. PAPS patients with history of cerebrovascular insults had elevated apo B concentrations in comparisons to the patients without this clinical finding.

Lower apolipoprotein AI levels and positive finding of anti-oxLDL antibodies were present in patients with history of venous thromboses.

Although a good correlation was found between Lp (a) and apo (a) concentrations in PAPS patients, cerebrovascular insults were only associated with elevated apo (a) concentrations. Logistic regression showed that the presence of elevated apo (a) concentrations is the most significant predictor of cerebrovascular insults in patients with PAPS.

Eighty percent of PAPS patients have either recurrent venous or arterial thromboses. The period between the initial occurrence and the recurrence may last several days or several years. Laboratory diagnosis of PAPS is currently based on persistently positive anticardiolipin (aCL) and/or lupus anticoagulant (LA) tests. Nevertheless, neither the aCL nor LA tests discriminate between patients with PAPS with diverse clinical manifestations. With the use of new appropriate laboratory tests, it may be possible to detect PAPS patients with diverse clinical manifestations and thus identify patients at increased risk.

In: Genetics Research Summaries
Editors: L. E. Hewitt and J. W. O'Neil

ISBN: 978-1-62257-646-3

Chapter 4

A Review of the Association of Apolipoprotein E and Delirium

Dimitrios Adamis[1,*] and Alastair J. D. Macdonald[2]
[1]Research and Academic Institute of Athens, Athens, Greece
[2]Institute of Psychiatry, London, UK

Research Summary

Introduction: Apolipoprotein E is a protein which has long been known as a major determinant of lipid transport and metabolism. The role of APOE in Alzheimer's disease, cognitive decline, and other dementias has been intensively investigated. However the role of APOE in delirium has only recently been investigated. In this review the relevant papers and the possible patho-physiological mechanisms relating APOE to delirium will be discussed.

Method: MEDLINE and other databases have been searched with the relevant keywords, and manually from the references of relevant papers. Only 5 papers were found which examined the relationship between APOE and delirium. Data were extracted from 4 of them and were pooled for further analysis. 2 of these studies explicitly examined the relationship between APOE and recovery of delirium.

Results: Only one study found a relationship between the APOE epsilon 4 allele and postoperative delirium while all the others reported no significant relationship. A meta-analysis from the 4 studies (pooled data from 692 participants) showed no heterogeneity (Q statistic, p=0.08) and the fixed model was used with the inverse variable weight. The possession of the APOE epsilon 4 allele has a small (OR:1.27, CI:0.87-1.86), non significant (p=0.22) effect in the presence or absence of delirium. Both studies which examine the relation of APOE and recovery of delirium found that APOE epsilon 4 carriers experienced a late or no recovery from delirium.

* Correspondence concerning this article should be addressed to: Dr. Dimitrios Adamis MSc, GradStat, MD Consultant in Old Age Psychiatry, Research and Academic Institute of Athens, 27 Themistokleous Street & Akadimias, Athens, 106 77, Greece. Tel/Fax: +30 210 33 040 43, email: dimaadamis@yahoo.com.

Conclusion: There is no association between APOE and the occurrence of delirium. However studies have been carried out in selective populations (65 years+) and their results may not be applicable to younger populations. Regarding the relationship between the APOE epsilon 4 allele with no recovery from delirium, both studies which found a positive relation lack power and further research is needed in this direction. Possible pathophysiological mechanisms are that the APOE epsilon 4 allele predisposes patients to uninhibited central nervous system inflammation during acute illness, with release of inflammatory mediators which could be important in the development of delirium. Current evidence suggests that apoE may modify central nervous system and systemic inflammatory responses in an isoform-specific manner. Apart from inflammatory mechanisms, APOE may contribute to brain dysfunction by blocking both nicotine and acetylcholine receptors causing the anticholinergic effect which is assumed in delirium. It is premature to suggest putative pathophysiological mechanisms. Confirmation and clarification in larger studies could have important clinical implications for predicting prognosis and for treatment of delirium.

In: Genetics Research Summaries
Editors: L. E. Hewitt and J. W. O'Neil
ISBN: 978-1-62257-646-3

Chapter 5

Apolipoproteins and Lipids in Overweight Men with Familial Hypertriglyceridemia

Judita Koprovičová and Jozef Kollár
Institute of Experimental Medicine,
Medical Faculty of the P.J. Šafárik University, Košice,
Slovak Republic

Research Summary

Background. Apolipoproteins (apo), mainly apoB_{100}, apoA-I, apoB/apoA-I ratio and apoC-III, are strong independent predictors of early atherosclerosis and initial coronary events. ApoC-III blood serum levels have been proposed to reflect triglyceride (TG) metabolism more accurately than the serum level of TG. Triglycerides, especially TG-rich lipoproteins (TG-RLPs) with apoC-III, may provide additional information on apoB_{100}.

Objectives. The aim of the pilot study was to estimate lipids and apolipoprotein serum levels in overweight subjects with familial hypertriglyceridemia (FHTG). The results were compared with the Control group (C) of the same age, sex, gender and profession.

Methods. The 23 overweight clerks (Ow) with FHTG and the 21 healthy subjects (Controls) with normal body weight, lipid and apolipoprotein serum levels were screened. These Ow subjects represent 3.2% from a group of 710 pre-screened clerks, which is a representative sample of the Slovak population. The mean level of the BMI of the Ow/controls was 27.3±1.7 kg.m^{-2}/24.2±0.7 kg.m^{-2}. The mean age for both groups was 47±9 years of age.

The serum levels of apoA-I, apoC-II, and apoC-III were determined by using radial immunoassay and apoB_{100} by electroimmunoassay. Serum levels of the triglycerides (TG), total cholesterol (TC) and HDL cholesterol (HDL-C) were detected by using biochemical sets. The LDL cholesterol (LDL-C), non-HDL cholesterol serum levels, the value of apoB_{100}/apoA-I and other ratios were calculated.

The data were tested by using ANOVA-1, ANOVA-2 and Pearson's correlation test, inclusive of the contingency tables. Familial hereditary disposition was studied through the genealogy.

Results. In the Ow in comparison with the C, the mean of the apoC-III serum levels was significantly increased: 15.48±4.83/4.42±1.08 $mg.dl^{-1}$ ($p<0.002$), inclusive of the apoC-II: 7.48±1.54/3.02±0.21 $mg.dl^{-1}$ ($p<0.001$) and $apoB_{100}$ 111.2±18.24/67.31±9.67 $mg.dl^{-1}$ ($p<0.001$). The mean of the apoA-I serum level was found to be significantly decreased: 120.02±24.13/141.49±10.51 $mg.dl^{-1}$ ($p<0.001$). The mean value of the $apoB_{100}$/apoA-I ratio was significantly increased: 0.93±0.13/0.48±0.06 ($p<0.001$).

The mean serum levels of the TG, TC, and LDL-C were significantly increased: 3.8±1.21/0.82±0.26, 6.5±0.51/4.03±0.35, and 3.91±1.17/2.23±0.34 $mmol.l^{-1}$ ($p<0.001$). The mean serum level of the HDL-C was significantly decreased: 1.23±0.31/1.30±0.27 $mmol.l^{-1}$ ($p<0.05$). The mean value of the ratios TC/HDL-C, LDL-C/HDL-C, and TG/HDL-C was 5.28±1.83/3.10±0.44, 3.18±1.38/1.72±0.41, and 3.21±1.21/0.63±0.1 ($p<0.001$). TG/apoC-II was significantly increased: 0.51±0.11/0.27±0.05 ($p<0.003$); and HDL-C/LDL-C decreased: 0.31±0.19/0.58±0.19 ($p<0.01$).

Conclusion. FHTG in the Ow subjects is associated with significantly elevated TG, apo C-III and $apoB_{100}$ and significantly increased TCH, LDL-CH and decreased apo A-I and HDL-C serum levels. To the cluster of risk factors belong more variables, e.g., significantly increased CL95% value of the TG/apoC-II and the $apoB_{100}$/apoA-I ratios. The criteria for metabolic syndrome from lipids are larger than are usually described. The significantly-increased serum levels of apoC-II act as a compensatory effect to stimulate lipoprotein lipase activity. In spite of that, the increased mean of the TG serum levels of about 393.5% may be caused by high secretion of TG-RLPs into circulation. The overweight subjects for dyslipidemic pattern, significant changes in lipoproteins and derived ratios are exposed to the cluster of risk factors that promote metabolic syndrome and accelerate development of cardiovascular diseases.

In: Genetics Research Summaries
Editors: L. E. Hewitt and J. W. O'Neil

ISBN: 978-1-62257-646-3

Chapter 6

Apolipoprotein E: Microbial Friend or Foe?

Ruth F. Itzhaki* and Matthew A. Wozniak
Faculty of Life Sciences, University of Manchester,
Manchester, UK

Abstract

Apolipoprotein E (ApoE) is a strikingly versatile protein. It plays a role not only in transport of lipids in the blood and the brain but also in the immune system and, as recent studies indicate, it appears to be involved too in the response of the host to diverse infectious agents – determining whether or not the host becomes infected, and if infected, whether or not the consequences are insignificant, mild, or severe.

The gene encoding apolipoprotein E (APOE) is polymorphic, with three main alleles (APOE-ε2, APOE-ε3, and APOE-ε4) encoding three protein isoforms (apoE2, apoE3, and apoE4). Possession of specific APOE alleles has been shown to determine the outcome of infection with several diverse pathogens, including herpes simplex virus type 1, varicella zoster virus, herpes simplex virus type 2, HIV, hepatitis C virus, *Mycobacterium tuberculosis* (an intracellular bacterium), and *Plasmodium falciparum*. Here we review these studies and describe possible mechanisms that underlie the involvement of apoE in determining the outcome of infection.

* Correspondence concerning this article should be addressed to: Professor Ruth Itzhaki, Faculty of Life Sciences (North Campus), Moffat Building, The University of Manchester, PO Box 88, Manchester M60 1QD, UK +44 (0) 161 306 3879; +44 (0) 161 306 3887; Email: ruth.itzhaki@manchester.ac.uk.

In: Genetics Research Summaries
Editors: L. E. Hewitt and J. W. O'Neil
ISBN: 978-1-62257-646-3

Chapter 7

Apolipoprotein AV: The Riddle of the Sphinx

Xian-sheng Huang, Shui-ping Zhao**, *Min Hu, Lin Bai and Qian Zhang
Department of Cardiology, the Second Xiangya Hospital of Central South University, Renmin Zhonglu, Changsha, Hunan, PR China

Research Summary

Hypertriglyceridemia has been acknowledged as an independent risk factor for coronary artery disease (CAD). Apolipoproteins (apo) play a determinant role in lipid homeostasis. Alterations in the levels of these functionally specialized proteins dramatically influence plasma lipid concentrations. Apolipoproteins that have been shown to affect plasma triglyceride levels include apolipoprotein E (apoE) and the C-apolipoproteins (apoCI, apoCII, and apoCIII).

The recently identified apoAV is the newest member of this family of apolipoproteins affecting plasma triglyceride levels. ApoAV gene (APOAV) was identified 30 kb proximal to the well-characterized APOAI/CIII/AIV gene cluster. ApoAV protein (apoAV) is a 39-kDa protein (343 amino acids) that is expressed exclusively by the liver. Plasma triglyceride concentrations in mice were 4-fold increased on deficiency of the endogenous APOAV gene and were decreased by 65% on expression of the human APOAV gene. Subsequent studies have shown that adenovirus-mediated transfer of murine APOAV to mice resulted in a dose-dependent reduction of plasma triglycerides. Similarly, genetic studies in human populations have identified several polymorphisms in APOAV that correlate with increased plasma triglycerides, familial combined hypertriglyceridemia and risk for cardiovascular disease. Furthermore, inherited APOAV deficiency is associated with severe hypertriglyceridemia in

* Correspondence concerning this article should be addressed to: Shui-ping Zhao, Department of Cardiology, The Second Xiangya Hospital, Central South University, No.139, Renmin Zhonglu, Changsha, Hunan, 410011, China. Fax: 86-0731-5360309; E-mail: zhaosp@medmail.com.cn.

humans. These studies have thus established an important role of apoAV in triglyceride homeostasis. However, somewhat paradoxically, its plasma concentrations, which are relatively low compared with other apolipoproteins, are elevated in patients with hypertriglyceridemia. Thus, although apoAV must be important in human triglyceride metabolism, aspects of its pathophysiological mechanistic roles still need to be better understood.

In: Genetics Research Summaries
Editors: L. E. Hewitt and J. W. O'Neil

ISBN: 978-1-62257-646-3

Chapter 8

Structure of Apoproteins of Hen Egg Yolk Low-density Lipoproteins in Relation to Functionalities: A Review

Pascale Jolivet[1,2,*], *Céline Boulard*[1,2], *Thierry Chardot*[1,2] *and Marc Anton*[3]
[1]Chimie Biologique, Thiverval-Grignon, France
[2]AgroParisTech, Chimie Biologique, Thiverval-Grignon, France
[3]Biopolymères, Interactions, Assemblages, Nantes, France

Research Summary

Apoproteins of low-density lipoproteins (LDL) contained in the hen egg yolk have been demonstrated as being essential to the emulsifying and gelling properties of yolk. The knowledge of their structure is necessary to better understand these functional properties. LDL were purified and their protein components were separated by SDS-PAGE. The analysis by liquid chromatography–tandem mass spectrometry of the peptides obtained from the trypsin digestion of protein gel bands led to the identification of two major proteins. Apovitellenin I was identified as a monomer and a dimer. Its amino acid sequence was totally confirmed and molecular mass determination by liquid chromatography–mass spectrometry showed that it did not present any post-translational modifications but only a slight heterogeneity by the loss of one or two amino acids at the C-terminal part of the protein. Apolipoprotein B (apo B) was identified by seven bands corresponding to eight protein fragments resulting from the maturation of the hen blood apo B protein through cathepsin D

* Correspondence concerning this article should be addressed to: Pascale Jolivet, UMR AgroParisTech/INRA 206, Chimie Biologique, Centre de Biotechnologie Agro-Industrielle, Thiverval- Grignon, 78850, France. Tel: 33 1 30 81 54 67. Fax: 33 1 30 81 53 73. E-mail address: jolivet@grignon.inra.fr.

proteolysis. The identity of the fragments was determined by the observation of the sequence coverage by trypsin peptides. Among these eight apoproteins, two protein fragments less accessible to proteases appeared to be firmly associated with lipids and could be enriched in β-sheets by homology with human blood apolipoprotein B-100 (apoB-100). This result combined with the sequence alignment with human apoB-100 has confirmed the pentapartite structure of hen apo B, which is composed of three amphipathic α-helical domains alternating with two amphipathic β-strand domains, NH_2-$\beta\alpha_1$-β_1-α_2-β_2-α_3-COOH.

In: Genetics Research Summaries
Editors: L. E. Hewitt and J. W. O'Neil
ISBN: 978-1-62257-646-3

Chapter 9

An Overview of ApoE for Application in Gene Therapy

Monica Rinaldi[1], *Sandra Iurescia*[1], *Daniela Fioretti*[1], *Emanuela Massi*[1,2] *and Emanuela Signori*[1,2,*]
[1]National Research Council (CNR), Institute of Neurobiology and Molecular Medicine, Area Ricerca Tor Vergata, Rome, Italy
[2]University Campus Bio-Medico of Rome, School of Medicine, Laboratory of Molecular Medicine and Biotechnology, Rome, Italy

Research Summary

Atherosclerosis, one of the most prevalent and severe age-related diseases, is the leading cause of death in industrialized countries and it is becoming an increasingly worldwide risk to health. As the size of elderly population grows, the incidence of atherosclerosis will increase, necessitating the exploration of radical, but potentially powerful, therapeutic strategies. One important goal is the availability of therapeutic molecules effective against the atherosclerotic plaque progression and able to stabilize advanced lesions. This generally requires robust measures to improve plasma lipoprotein profiles. Pharmaceutical agents currently available do not support significant changes in plasma lipoprotein levels. So that, strong new agents and therapies are needed in order to treat hyperlipidemia and atherosclerosis.

Apolipoprotein E (ApoE), a key regulator in cholesterol-rich lipoprotein metabolism, is considered a strong candidate for treating hypercholesterolemia and cardiovascular disease. ApoE gene therapy could be proposed as a new valid approach for the correction of hyperlipidemic conditions and atherosclerosis by using viral and non-viral vectors. In this review we discuss the major developments in both viral and non-viral gene transfer systems

* Correspondence concerning this article should be addressed to: Dr. Emanuela Signori, PhD CNR - Department of Medicine, Laboratory of Molecular Pathology and Experimental OncologyInstitute of Neurobiology and Molecular Medicine (INMM), Area Ricerca Tor Vergata, Via Fosso del Cavaliere 100, 00133 Rome, Italy. Phone: +39 064993 4232; Fax: +39 064993 4257; e-mail: emanuela.signori@artov.inmm.cnr.it.

aimed at achieving efficient protocols for treating hypercholesterolemia and atherosclerosis by using ApoE as therapeutic gene.

Key scientific advances on the ability of ApoE gene transfer to markedly improve outcome after cerebral ischemia, suggesting a potential strategy in human stroke therapy, are also described in this review.

The characterization of ApoE and its many functions has provided insight into the ultimate potential of this protein as a possible therapeutic agent for the treatment of several diseases. Possible therapeutic strategies that encompass the use of ApoE in gene therapy for Alzheimer's disease and other central nervous system (CNS) diseases will be also discussed.

In: Genetics Research Summaries
Editors: L. E. Hewitt and J. W. O'Neil
ISBN: 978-1-62257-646-3

Chapter 10

The Heritabilities, Genetics and Networks of Apolipoproteins

Mogens Fenger
Department of Clinical Biochemisitry and Molecular Biology,
University Hospital of Copenhagen, Hvidovre, Denmark
The Faculty of Health, University of Copenhagen, Denmark

Research Summary

The interest in apolipoproteins is nourished by their importance in lipid metabolism and their relation to common disease as obesity, diabetes mellitus, and atherosclerosis. The assembling of apolipoproteins with lipids forms the lipoproteins, the main structures being very low density lipoproteins, low density lipoproteins and high density lipoproteins. The lipoproteins are vehicles for transport of lipids ingested or synthesized as nutrients to all tissues, and to remove debris from tissue metabolism to be processed mainly in the liver. However, the lipoproteins are not simply transporters of lipids, but are integrated in a complex network of metabolic processes, including exchange of lipids between the lipoproteins and participates in regulating of lipid metabolism *per se*. The latter is accomplished by modifications of lipids and their distribution within the lipoproteins and by interactions with receptors regulating cellular activities. The lipoproteins thus forms an extracellular, metabolic sub-network of the lipid metabolism. This sub-network is engaged in a two-way communicates with cellular processes through lipoprotein receptors, thereby delivering lipids and other metabolites as nutrients and hormonal regulators to the cells, which in turn regulates the production of apolipoproteins.

The genetic variations of the apolipoproteins and the network they operate in would predict an almost infinitive number of phenotypic expressions of the transport system. Although this diversity is the consequence of and the foundation of evolutionary precesses, the viability of a species requires robustness of the system. This implies, that it is the fitness of the network as a functional module that is deterministic, not the single genes in as much that all genes function in a context, and consequently, the number of permissible network variations is limited. The lipoprotein network is particular supportive of this scenario, as many of the apoplipoproteins are segregated *en block*, suggesting that specific multi-gene haplotypes are inherited, limiting the number of admissible genetic variations in the network. The transport network is further constrained, as the components do not interact randomly. Nevertheless, the genetics and functionality of the network is still very complex. The heterogeneity of the functionality of apolipoproteins and lipoproteins must therefor be modelled as integrated metabolic-genetic networks if we shall have any hope to describe and define the architecture of this particular metabolic sub-system.

In: Genetics Research Summaries
Editors: L. E. Hewitt and J. W. O'Neil
ISBN: 978-1-62257-646-3

Chapter 11

Apolipoprotein E Gene Polymorphism and Metabolic Syndrome in a Free-living Elderly Community

Maria Gabriela Valle Gottlieb*[1], *Carla Helena Augustin Schwanke*[1], *Jacqueline da Costa Escobar Piccoli*[2], *Rodolfo Herberto Schneider*[1], *Luiz Carlos Bodanese*[1], *Euler Esteves Ribeiro*[3] and *Ivana Beatrice Mânica da Cruz*[4]

[1]Pontifícia Universidade Católica do Rio Grande do Sul-PUCRS, Brazil
[2]Universidade Integrada do Alto Uruguai – Santo Ângelo, Brazil
[3]Universidade do Estado do Amazonas, Brazil
[4]Universidade Federal de Santa Maria, Brazil

Abstract

Introduction: apolipoprotein E (apoE) plays a key role in lipid and lipoprotein metabolism. The apoE gene shows three common polymorphic alleles: ε2, ε3 and ε4, and six genotypes ε2/2, ε2/3, ε2/4, ε3/3, ε3/4 and ε4/4. Previous studies have suggested that apoE is a "frailty gene" since the ε4 allele is associated with several morbidities including cardiovascular diseases and Alzheimer's disease. Despite the results obtained to date, few investigations have been performed to test if ε4 could be associated with MS. Metabolic syndrome (MS) is defined as a set of cardiovascular risk factors with complex metabolic-cellular alterations. There is also a strong gene-environmental component. The etiology of MS is still unknown; however, it is associated with several lipid and lipoprotein disorders.

Objective: The aim of this study was to determine the association between apoE polymorphism and MS in an elderly population in southern Brazil.

Methods: A case-control study was designed including Caucasian participants who were previously investigated in a epidemiological project that involved the study of gene-environmental interactions in aging and related diseases conducted in the South Brazil region (Gravataí, Rio Grande do Sul State). Out of 1158 subjects, we selected a sub-sample (controls and MS cases).The diagnostic criteria of MS was performed by the Third Report of the U. S. National Cholesterol Education Program Adult Treatment Panel (NCEP-ATP III). apoE genotypes were identified by PCR-RFLP using genomic DNA.

Results: A total of 366 subjects were investigated (206 controls and 106 with MS). The mean age was 67.0±7.2 in control and 67.0±6.2 in MS subjects, respectively. The apoE genotype frequencies were: ε3/ε3=60.7%, ε3/ε4= 24.6%, ε3/ε2= 9.6 %, ε4/ε4= 2.2% and ε4/ε2= 3.0%. We found significant association between ≥ 150mg/dL triglycerides and apoE polymorphism in MS group (p=0.021). MS ε4ε4 carriers presented significant lower frequency in subjects with ≥ 150 mg/dL triglycerides than other allele carriers. We found significant association between higher ε4 allele frequency and control group (p=0.031). In MS group ε4 allele frequency was 21.7% (n=23) whereas in control group was 33.1% (n=86). Odds ratio was 0.561 (95% Confidence Interval= 0.330-0.952) to ε4 carriers to present MS when compared to other alleles carriers. Additional logistic regression analysis showed that association was independent of sex, age, diabetes and hypertension (p=0.029). However, it is important that additional studies to be performed to elucidate if the MS and apoE polymorphism present a real association.

In: Genetics Research Summaries
Editors: L. E. Hewitt and J. W. O'Neil
ISBN: 978-1-62257-646-3

Chapter 12

Why is apoE Genotype a Risk Factor for Multiple Chronic Neurological Disorders?

Britto P. Nathan[1] and Robert G. Struble[2]
[1]Department of Biological Sciences, Eastern Illinois University, IL, US
[2]Center for Alzheimer's Disease and Related Disorders,
Southern Illinois School of Medicine, Springfield, IL, US

Research Summary

We propose that the primary role of apolipoprotein E (apoE) in the brain is to support repair and remodeling after neuronal injury. Epidemiological studies have documented that inheriting one or more apoE4 allele increased the risk, exacerbated the severity and accelerated the clinical progression of several chronic neurological conditions and diseases including Alzheimer's disease, Parkinson's disease, and brain trauma. Data suggest that apoE3 is more efficient in supporting neuronal repair than is apoE4. Therefore, our working hypothesis is that the involvement of apoE4 in numerous chronic neurological diseases does not necessarily represent a disease-specific effect on risk, but rather less efficient repair processes. Poorer repair results in earlier or more severe manifestations of the disease. Understanding the mechanisms whereby apoE isoforms differentially modify neuronal repair and remodeling will lead to strategies to delay or arrest disease progression.

In: Genetics Research Summaries
Editors: L. E. Hewitt and J. W. O'Neil
ISBN: 978-1-62257-646-3

Chapter 13

The Relationship of Postprandial Lipemia and Apolipoprotein Composition of High Density Lipoproteins

O. V. Alexandrovitch*[*]*, I. N. Ozerova, V. A. Metelskaya and N. V. Perova
National Research Center for Preventive Medicine,
Moscow, Russia

Research Summary

High density lipoprotein (HDL) compose of two type particles: lipoproteins containing only apolipoprotein (apo) AI (LpAI), and those containing both apo AI and AII (LpAI:AII). Apo AI has been widely accepted to play antiatherogenic role, whereas function of apo AII is less clear. Apo AII has been suggested to regulate metabolism both of triglyceride-rich lipoproteins and glucose.

This study aimed to investigate whether postprandial lipemia (elevation of serum triglycerides level after single fat load) depends on apo AII/AI molar ratio, which reflects the proportion of two HDL particles types.

Analyses were carried out after overnight fasting and 6 hours after single fat load as a postprandial state model.

Thirty-one subjects (15 males/16 females) aged 20-66 with fasting normoglycemia were assigned into two groups according to 50th percentile of molar apo AII/AI ratio: <0,37 – group 1 (n=15), and ≥0,37 – group 2 (n=16). No significant differences in age and body mass

[*] Correspondence concerning this article should be addressed to: O. V. Alexandrovitch, National Research Center for Preventive Medicine, 10, Petroverigsky lane, 101990 Moscow, Russia. E-mail: oalexandrovitch @gnicpm.ru.

index between groups were found. Subjects from group 2 as compared with group 1 had lower apo AI (1,37 vs 1,57 g/l, p=0,025), HDL cholesterol (0,97 vs 1,40 mmol/l; p=0,009), HDL phospholipids (1,08 vs 1,23 g/l, p=0,024) levels, but elevated triglycerides (1,90 vs 1,00 mmol/l, p=0,053), apo AII (0,35 vs 0,30 g/l; p=0,06) levels. Moreover, higher insulin level (13,8 vs 3,4 μU/ml; p=0,005) and HOMA-IR (1,7 vs 0,4; p=0,006) were found in group 2, suggesting latent insulin resistance. No postprandial changes were detected in serum total cholesterol and apo AII concentrations in both groups. After 6-h of single fat load no significant differences between groups were obtained in incremental areas under the curves (AUICs) for apo AI, apo B, HDL cholesterol, and HDL phospholipids. At the same time, postprandial triglycerides response was higher (5,18 vs 3,23 mmol/l x 6h; p=0,01) and LDL cholesterol response was lower (-1,86 vs - 0,89 mmol/l x 6h) in group 2.

In conclusion, the more pronounced postprandial triglycerides response obtained in subjects with fasting normoglycemia and higher apo AII/AI molar ratio may be due to increased portion of LpAI:AII particles within HDL spectrum, because apo AII is supposed to impair the hydrolysis of triglyceride-rich lipoproteins thus inducing postprandial hypertriglyceridemia.

In: Genetics Research Summaries
Editors: L. E. Hewitt and J. W. O'Neil
ISBN: 978-1-62257-646-3

Chapter 14

The Role of Apoproteins in Modern Cardiovascular and Medical Research

Georgios I. Tagarakis
Cardiothoracic Department, University Hospital of Thessaly,
Larissa, Greece

Research Summary

The term "Apoprotein" derives from the greek language and consists of the words protein (πρωτεΐνη), with the known, commonly used meaning referring to polypeptide molecules and the preface "apo-από", which in this context has the meaning of a complementing, that is of a second consisting or prosthetic part. In other words, apoproteins are the union of proteins with other molecules, which in terms of modern biochemistry could be either apoenzymes or apolipoproteins. The latter, bearing a strong relationship with the cholesterol and lipid metabolism in general, stand in the focus of contemporary research, as they are connected in multiple ways to the aetiology of all forms of cardiovascular disease, but also to the genesis of Alzheimer's disease and Other Dementias.

In this short article we will deal with the apoprotein molecules (from the total six groups of the category) that mainly have come under the loop of modern science, trying to show their potential or proven usefulness for the prevention, diagnosis and therapy of various nosological entities.

In: Genetics Research Summaries
Editors: L. E. Hewitt and J. W. O'Neil

ISBN: 978-1-62257-646-3

Chapter 15

Codon and Codon Pairs Usage in Bacteria

Boris I. Bachvarov, Kiril T. Kirilov and Ivan G. Ivanov*
Institute of Molecular Biology,
Bulgarian Academy of Sciences, Bulgaria

Research Summary

The new generation of fast and low-cost DNA sequencing methods led to a rapid accumulation of genomic data in the DNA databases after the year 2000. Hundreds of bacterial genomes have been sequenced during the last decade, however, in rare cases they were profoundly analyzed. Apparently it is much easier today to generate raw DNA sequencing data rather than to decipher their biological meaning. This study aims to shed light on the relationship between codon usage and gene expression in 158 prokaryotes belonging to the two major classes: *Archaea* and *Bacteria*. *Archaea* includes 2 families (*Crenarchaeota* and *Euryarchaeota*) and *Bacteria* 13 families (*Actinobacteria, Aquificae, Bacteroidetes, Chlamydiae, Chlorobi, Cyanobacteria, Deinococcus-Thermus, Firmicutes, Fusobacteria, Planctomycetes, Proteobacteria, Spirochaetes* and *Thermotogae*) and the total number of genera was 9 for *Archaea* and 22 for *Bacteria*. In order to study the link between codon usage and gene expression four local DNA databases were created containing respectively: a) all protein coding sequences of each organism; b) highly expressed genes; c) ribosomal proteins genes and d) low expressed genes. These databases were used to compare codon usage patterns of: a) different strains of one bacterial species; b) different subsets of genes in one species and c) different bacterial genomes. They were also employed to identify missing and atypical codons as well as to search for correlation between codon usage and the formal prokaryotic taxonomy. The obtained results indicate that the formal taxonomy of many bacterial species does not fit their codon usage pattern (as predicted by the "Genome Theory"). On the other hand there are obvious similarities in the codon usage pattern of

* Corresponding author: e-mail: iivanov@bio21.bas.bg

distant (according to the formal taxonomy) organisms. The latter suggests that either the Genome Theory or the official taxonomy of microorganisms need revisions. Taking into consideration that at least two tRNAs occupy the two (A and P) functional ribosomal sites during the translation elongation step, we have also studied the effect of neighbor codons (codon pairs) on gene expression. Aiming to identify preferential and rare codon pairs, we have developed own criteria for determining their frequency of occurrence and for evaluation of their possible effect on translation. This approach was applied to analyze the distribution of all (3904) codon pairs in the *Escherichia coli* genome and we found that their frequency of occurrence varied from zero to 4913 times. The predicted effect of some (particularly 3'-terminal) codon pairs on gene expression is confirmed experimentally.

In: Genetics Research Summaries
Editors: L. E. Hewitt and J. W. O'Neil

ISBN: 978-1-62257-646-3

Chapter 16

Maillard Reaction and Spontaneous Mutagenesis in *Escherichia coli*

***Roumyana Mironova*[1], *Yordan Handzhiyski*[1], *Toshimitsu Niwa*[2], *Alfredo Berzal-Herranz*[3], *Kirill A. Datsenko*[4], *Barry L. Wanner*[4] *and Ivan Ivanov*[1]**

[1]Department of Gene Regulations, Institute of Molecular Biology, Bulgarian Academy of Sciences, Sofia, Bulgaria
[2]Department of Clinical Preventive Medicine, Nagoya University School of Medicine, Nagoya, Japan
[3]Instituto de Parasitología y Biomedicina 'López-Neyra', CSIC, Parque Tecnológico de Ciencias de la Salud, Armilla, Granada, Spain
[4]Department of Biological Sciences, Purdue University, West Lafayette, IN, US

Research Summary

DNA is a fragile organic molecule of finite chemical stability, capable of interacting with a variety of chemicals. In the context of the living cell this DNA reactivity is manifested as mutability. In addition to the chemical mutability, the perpetuation of DNA via imperfect replication is another source of DNA changeability. It is worth mentioning that DNA repair is error-tolerant and always leaves behind some lesions exist. Were repair mechanisms perfect, then the door of evolution would be locked. Risky rather than benign at individual level, mutations are the living things' long term strategy for survival and evolution. The creative power of mutations can explain why organisms have evolved special mechanisms to enhance mutagenesis under certain conditions. Hypermutation in bacteria under stress (SOS response) as well as hypermutation of immunoglobulin genes in response to antigen challenge in vertebrates are key examples.

The present study is focused on spontaneous mutagenesis in *Escherichia coli* K-12.

Spontaneous mutations are the net result of DNA damage and repair under normal physiological conditions. Bacterial DNA is loosely packed with proteins and occupies almost the entire cellular space. Therefore, it is extremely vulnerable to chemical attack, especially during replication and transcription, when the two DNA strands are separated. In the same arena, the cytoplasm, and at the same time, play also many small but highly reactive compounds. Although the collision of the latter with DNA seems to be unavoidable, to date the impact of ordinary cellular metabolites on spontaneous mutagenesis has not been systematically studied.

In year 2012 we shall celebrate the 100 anniversary of the discovery of the carbonyl-amine reaction by the French chemist Louis Camille Maillard (Maillard, 1912). Here, we provide evidence that the Maillard reaction, referred to as non-enzymatic glycosylation or glycation in the literature, is an important endogenous source of spontaneous mutations in *E. coli*. Experimental data allowed us to estimate that glycation affects on average one per 10^5 to 10^4 nucleotides in the *E. coli* chromosome. Based on experimental evidence, we postulate the existence in *E. coli* of a novel DNA repair enzyme, DNA amadoriase or DNA deglycase, especially designed to combat glycation induced spontaneous mutagenesis in *E. coli*.

In: Genetics Research Summaries
Editors: L. E. Hewitt and J. W. O'Neil

ISBN: 978-1-62257-646-3

Chapter 17

Nucleoid Architecture and Dynamics in Bacteria

Ryosuke L. Ohniwa[*1,2], *Kazuya Morikawa*[2], *Toshiko Ohta*[2], *Chieko Wada*[1,3] and *Kunio Takeyasu*[1]

[1]Laboratory of Plasma Membrane and Nuclear Signaling, Kyoto University Graduate School of Biostudies, Yoshida-Konoe, Sakyo-ku, Kyoto, Japan

[2]Institute of Basic Medical Sciences, Graduate School of Comprehensive Human studies, University of Tsukuba, Tennodai, Tsukuba, Japan

[3]Yoshida Biological Laboratory, Yamashina, Kyoto, Japan

Research Summary

Both prokaryotes and eukaryotes store their genomic DNA in an environment, which balances the physical properties of double-stranded DNA with the mechanical effects of DNA-protein interactions. The origin of genome packing remains a mystery; it seems to go back to the very beginning of life itself. Since then, the principle of higher-order genome construction and architecture has been maintained and shared among three domains of living things. The genomes are stored as flexible higher-order structures that can shuttle between relatively active and inactive states. The major differences reside in the structural protein components of the architecture. Bacteria and Eukaria utilize HU and histones as the most fundamental structural proteins, respectively, forming the nucleoid in bacterial cell and the nucleosomes in the nucleus. On the other hand Archaea employ either HU or histones depending on their phylogenetic origins. Nevertheless, the hierarchies of nucleoid fibers look extremely similar among different species in different domains. Focusing upon the mechanism of bacterial genome folding, recent structural investigations have revealed a step-

* Correspond to: R.L. Ohniwa, Institute of Basic Medical Sciences, Graduate School of Comprehensive Human Sciences, University of Tsukuba, Tsukuba, 305-8575 Japan Phone and Fax: +81-29-853-3928; e-mail: ohniwa@md.tsukuba.ac.jp

wise folding of the genome DNA; from 10 nm to 30 nm fibers, and to 80 nm and further condensed beaded structures, depending upon the growth conditions and differential contribution of nascent single-stranded RNA. The nucleoid condensation is mainly brought about by another nucleoid protein, Dps, which occurs widely in the bacterial domain, but not in Archaea or Eukarya. The regulatory mechanisms of the *dps* gene are greatly different among bacterial species. In γ-Proteobacteria such as *Escherichia coli*, the *dps* gene is governed by two regulatory systems, IHF-σs and OxyR, and is induced towards the stationary phase and under oxidative stress. The DNA topology control by *E. coli* Fis also participates in the regulation of the nucleoid condensation. In contrast, in Firmicutes such as *Staphylococcus aureus*, the gene is mainly controlled by oxidative stress-response system, PerR, and the Dps expression is directly coupled with the nucleoid condensation. Here the key features of the nucleoid dynamics will be discussed from the biochemical and structural biological points of views.

In: Genetics Research Summaries
Editors: L. E. Hewitt and J. W. O'Neil
ISBN: 978-1-62257-646-3

Chapter 18

Genome Reduction in *Bacillus subtilis* and Enhanced Productivities of Recombinant Proteins

Yasushi Kageyama[1], Takuya Morimoto[1,2], Katsutoshi Ara[1], Katsuya Ozaki[1] and Naotake Ogasawara[2]

[1]Biological Science Laboratories, Kao Corporation,
Akabane, Ichikai, Haga, Tochigi, Japan
[2]School of Information Science,
Nara Institute of Science and Technology,Takayama,
Ikoma, Nara, Japan

Research Summary

Bacillus subtilis, which is one of the most important host microorganisms for large-scale industrial production of useful proteins, has a genome of 4.2 Mb with approximately 4106 protein-coding genes. Some of these genes are expected to be unnecessary for industrial production of proteins under controlled conditions and may be wasteful with regard to energy consumption. We attempted to reduce the genome size of *B. subtilis* by deleting unnecessary regions of the genome to allow the construction of simplified host cells as a platform for the further development of novel genetic systems with increased productivity.

First, we generated the strain MGB469 with deletion of all prophage (SPβ and PBSX) and prophage-like (pro1-7 and skin) sequences, with the exception of pro7, as well as two large operons that produce secondary metabolites (*pks* and *pps*). These cells showed normal growth, but no beneficial effects were observed with regard to recombinant protein production from plasmids carrying the corresponding genes. Second, we constructed several multiple-deletion mutants containing additional deletions in the MGB469 genome, resulting in total genome size reductions of 0.78 to 0.99 Mb. In most of the multiple-deletion series, extensive deletion mutants showed no beneficial improvements in traits as host strains. The strain MG1M with a total genome size reduction of 0.99 Mb showed unstable phenotypes with regard to growth rate, cell morphology, and recombinant protein productivity after successive culture, making it inappropriate for further studies. In addition, strain MGB943 derived from another lineage with genome reduction of 0.94 Mb showed reduced recombinant cellulase productivity. Finally, we generated another multiple-deletion series including the mutant MGB874 with a total genome deletion of 0.87 Mb. In comparison to wild-type cells, the metabolic network of the mutant strain was reorganized after entry into the transition state due to the synergistic effects of multiple deletions. Moreover, the levels of production of extracellular cellulase and protease from transformed plasmids carrying the corresponding genes were markedly increased.

Our results demonstrated the effectiveness of a synthetic genomic approach with reduction of genome size to generate novel and useful bacteria for industrial uses.

In: Genetics Research Summaries
Editors: L. E. Hewitt and J. W. O'Neil
ISBN: 978-1-62257-646-3

Chapter 19

Molecular Detection and Characterization of Food and Waterborne Viruses

***Alain Houde*[1], *Kirsten Mattison*[2], *Pierre Ward*[1] and *Daniel Plante*[3]**
[1]Agriculture and Agri-Food Canada, Food Research and Development Centre, St-Hyacinthe, Quebec, Canada
[2]Bureau of Microbial Hazards, Health Products and Food Branch, Food Directorate, Health Canada, Ottawa, Ontario, Canada
[3]Health Canada, Quebec Region, Longueuil, Quebec, Canada

Research Summary

Enterically transmitted viruses such as norovirus (NoV), rotavirus (RV), hepatitis A virus (HAV) and hepatitis E virus (HEV) are currently recognized as a major cause of food and waterborne illness in humans worldwide. Most of these viruses cannot be cultured *in vitro*, are stable in the environment for long periods of time, are generally infectious at low doses (only a few viral particles are needed to trigger illness) and would be usually present in low numbers in contaminated food and water samples. There are also increasing concerns about the possible zoonotic transmission or the emergence of new recombinant strains from animal enteric viruses closely related to human pathogenic strains. Until recently, the viral contamination of food and water remained undetected due to a lack of appropriate detection methods. Electronic microscopy (EM) and ELISA assays commonly used on stool suspensions were rather insensitive for food and water applications. Due to their increased sensitivity, conventional and now real-time quantitative reverse transcription PCR (qRT-PCR) assays were developed and widely used for the detection and identification of the nucleic acid of these challenging viruses. Along with microarrays, these new approaches have also allowed significant advances in the molecular typing of these organisms. Rapid, sensitive and reliable molecular detection and typing methods will have a significant impact in the mitigation of outbreaks and could contribute in the prevention of food/waterborne transmission. This review will provide recent trends and advances in the field of DNA polymerase-based detection and characterization methods of these 4 main food and waterborne viruses. It will focus on the use of different conventional and qRT-PCR methods and the microarray technology.

In: Genetics Research Summaries
Editors: L. E. Hewitt and J. W. O'Neil
ISBN: 978-1-62257-646-3

Chapter 20

hTERT in Cancer Chemotherapy: A Novel Target of Histone Deacetylase Inhibitors

Jun Murakami[1], Jun-ichi Asaumi*[*1], Hidetsugu Tsujigiwa[2], Masao Yamada[2], Susumu Kokeguchi[3], Hitoshi Nagatsuka[4], Tatsuo Yamamoto[5] and You-Jin Lee[6]

[1]Department of Oral and Maxillofacial Radiology,
[2]Department of Virology,
[3]Department of Oral Microbiology,
[4]Department of Oral Pathology,
[5]Department of Preventive Dentistry,
Graduate Schools of Medicine, Dentistry and Pharmaceutical Sciences,
Okayama University, Okayama, Japan
[6]Jeonnam Biotechnology Research Center, Korea

Research Summary

Chromatin structure plays an important role in the regulation of gene transcription. Chromatin structure can be modified by various post-translational modifications, including histone acetylation, phosphorylation, methylation and ribosylation. Among those modifications, histone acetylation/deacetylation is the most important mechanism for regulating transcription and is regulated by a group of enzymes known as histone acetyltransferases/histone deacetylases (HDACs).

Recently, HDAC inhibitors have been shown to be a novel and promising new class of anti-cancer agent that can regulate the transcription of genes by disrupting the balance of acetylation/deacetylation in particular regions of chromatin. A number of HDAC inhibitors

[*] E-mail: asaumi@md.okayama-u.ac.jp and jun-m@md.okayama-u.ac.jp, Telephone number: +81-86-235-6621, Fax number: +81-86-235-6709

are currently in phase I and II clinical trials against a variety of cancers. Although some promising candidates have been identified (e.g., $p21^{WAF1}$ and c-Myc), the precise molecular targets remain uncertain. In this article, we focus on one of the DNA polymerases, telomerase, as a new candidate molecular target for HDAC inhibitors. Telomerase is composed primarily of the catalytic subunit (hTERT) and the RNA template (hTERC), and its activity correlates with levels of hTERT mRNA. hTERT expression is apparently governed by complicated regulatory pathways. Based on recent studies, the hTERT gene is likely to be targeted by histone acetylation/deacetylation.

In: Genetics Research Summaries
Editors: L. E. Hewitt and J. W. O'Neil
ISBN: 978-1-62257-646-3

Chapter 21

Dynamics of DNA Polymerase

Ping Xie *

Laboratory of Soft Matter Physics, Institute of Physics,
Chinese Academy of Sciences, Beijing, China

Research Summary

Replicative DNA polymerase (DNAP) is an enzyme that synthesizes a new DNA strand on a single-stranded template with a high processivity and a high fidelity. The high fidelity is mainly realized via a mechanism of proofreading that is performed at the exonuclease active site spatially separate from the polymerase active site. Here we present a detailed account of our proposed model for the processive nucleotide incorporation and switching transition of DNA between the polymerase and exonuclease active sites by DNAP. Based on the model, we present detailed theoretical studies on its dynamics. The moving time of DNAP to next site after a correct incorporation and that after an incorrect incorporation are analytically studied. The experimentally measured dependence of polymerization rate on tension applied to the template is well simulated. The transfer rates of DNA from the polymerase to exonuclease active sites after a correct incorporation and after incorrect incorporations as well as the transfer rate from the exonuclease to polymerase active sites are analytically studied. Moreover, the backward motion of DNAP when large tensions are applied to the template is also analytically studied. The theoretical results are in good agreement with available experimental data. Some predicted results are presented.

* E-mail address: pxie@aphy.iphy.ac.cn

In: Genetics Research Summaries
Editors: L. E. Hewitt and J. W. O'Neil

ISBN: 978-1-62257-646-3

Chapter 22

Inhibiting Viral DNA Polymerase for HBV Therapy

Min Jiang and Yuanhao Li
Vertex Pharmaceutical Inc., Cambridge, MA, US

Research Summary

Hepatitis B virus infection is a major public health problem worldwide especially in East Asia. The viral infection, if persistent, may lead to chronic hepatitis, cirrhosis, and hepatocellular carcinoma. In1998, FDA approved GlascoSmithKline's Epivir-HBV (Lamivudine), a small molecule drug that inhibits HBV DNA polymerase and interferes with viral replication. Since then, a few similar drugs have been approved and many candidates are currently at preclinical and clinical development. The use of small molecular drugs that target HBV DNA polymerase is a milestone in the treatment of chronic hepatitis B. In this review, we will focus on HBV DNA polymerase and discuss the preclinical and clinical development process for the HBV polymerase inhibitors. The DNA polymerase mutants associated with drug resistance will also be discussed. The mechanism of the drug resistance and further understanding of these DNA polymerase inhibitors may help the determination of better clinical regimen for HBV therapy and patient care.

In: Genetics Research Summaries
Editors: L. E. Hewitt and J. W. O'Neil
ISBN: 978-1-62257-646-3

Chapter 23

Modeling DNA Translocation and Unwinding by Helicase RecG

Ping Xie *
Laboratory of Soft Matter Physics, Institute of Physics,
Chinese Academy of Sciences, Beijing, China

Research Summary

RecG is a DNA helicase involved in the repair of damage at a replication fork by catalyzing the reversal of the fork to create a Holliday junction. Here, based on previous structural and biochemical studies a model is presented on how RecG catalyzes the processive translocation and unwinding of DNA so as to realize the fork reversal. In the model, a power stroke induces the DNA translocation and unwinding. Moreover, in order to effectively unwind DNA of a helical structure, it is assumed that the residues (i.e., the long α helix) connecting the wedge domain and the helicase domains in which the dsDNA-binding loop is located behave elastically along the torsional direction. Thus, accompanying the power stroke, the wedge domain is forced elastically to rotate relative to the dsDNA-binding loop that binds dsDNA strongly. Using the model, the calculated DNA translocation size per ATP hydrolysis is consistent with previous experimental data. Moreover, it is showed that RecG has a strong preference for negatively supercoiled DNA, i.e., RecG shows a much higher ATPase activity and longer processivity when interacting with the negatively supercoiled DNA than with the relaxed and/or positively supercoiled DNAs. This is also in agreement with the previous experimental data. In addition, the rotational rate or the rotational angle of RecG relative to DNA per ATP hydrolysis and the torsional elastic coefficient of the residues connecting the wedge domain and the helicase domains are predicted.

* E-mail address: pxie@aphy.iphy.ac.cn

In: Genetics Research Summaries
Editors: L. E. Hewitt and J. W. O'Neil
ISBN: 978-1-62257-646-3

Chapter 24

De Novo DNA Synthesis by DNA Polymerase

Xingguo Liang*[*], *Tomohiro Kato and Hiroyuki Asanuma
Department of Molecular Design and Engineering,
Graduate School of Engineering, Nagoya University,
Chikusa, Nagoya, Japan

Research Summary

In this chapter, the *de novo* DNA synthesis by DNA polymerase in the absence of any added template and/or primer is described. As early as the 1960s, Kornberg et al. reported that DNA polymerase I could *de novo* synthesize DNA, although it was later pointed out that the contaminated DNA or RNA may provide the seeds for DNA synthesis and the contaminated transferase that can polymerize dNDPs might play the role of template-independent DNA synthesis. The synthesized DNA polymers were characterized as homopolymer pairs such as poly(dA)/poly(dT) and poly(dI)/poly(dC), or the alternating copolymers poly(dA-dT), poly(dG-dC) and poly(dI-dC). In late 1990s, after a long silent time, Ogata et al. found that the highly purified thermophilic DNA polymerase could carry out *de novo* DNA synthesis at higher temperatures (> 65°C) and the synthesized DNA polymers were composed of repetitive palindromic sequences such as $(TACATGTA)_n$, and $(TAAT)_n$. Surprisingly, Liang et al. reported that thermophilic DNA polymerases could carry out *de novo* DNA synthesis even at room temperature. Basically, *de novo* DNA synthesis is considered to occur in two stages, although the detail mechanism is not clear. In the initial stage, dNTPs are polymerized to short DNA oligos, and then the selected repetitive sequences are elongated to long DNA in the elongation stage through self-priming and primer extension. More interestingly, the *de novo* DNA synthesis can be greatly accelerated by adding endonuclease such as restriction enzymes, which digest DNA at their recognition sites. A Cut-grow mechanism was proposed to explain both the paradox of endonuclease and the extremely high efficiency: the synthesized long DNA is cut to shorter seeds for the elongation of the next cycle so that DNA can be exponentially amplified. The possible biological and evolutionary roles of *de novo* DNA synthesis are also discussed.

In: Genetics Research Summaries
Editors: L. E. Hewitt and J. W. O'Neil
ISBN: 978-1-62257-646-3

Chapter 25

Bacteriophage φ29 DNA Polymerase: An Outstanding Replicase

Miguel de Vega* and Margarita Salas
Instituto de Biología Molecular "Eladio Viñuela" (CSIC),
Centro de Biología Molecular "Severo Ochoa" (CSIC-UAM),
Campus Universidad Autónoma, Cantoblanco,
Madrid, Spain

Research Summary

Due to the limited processivity of replicative DNA polymerases (replicases), as well as to their incapacity to unwind parental duplex DNA to allow replication fork progression, their replication efficiency depends on the functional assistance of accessory proteins as processivity factors and helicases. In addition, the inability of DNA polymerases to start de novo DNA synthesis requires the use of a short RNA/DNA molecule to provide the 3'-OH group required to initiate DNA replication. This requisite for a primer creates a dilemma to replicate the ends of linear genomes: once the last primer for the lagging strand synthesis is removed, a portion of ssDNA at the end of the genome will remain uncopied. Bacteriophage φ29 has overcome these issues by means of the unique catalytic features of an outstanding enzyme, the φ29 DNA polymerase. This replicase belongs to the family B (eukaryotic-type) of DNA-dependent DNA polymerases and has served as model to understand the enzymology of these polymerases. As most of the family B members, φ29 DNA polymerase contains both 3´-5´ exonuclease and polymerization activities residing in two structurally independent domains. During two decades, site-directed mutagenesis studies of individual residues contained in regions of high amino acid similarity have provided the functional insights of this enzyme, extrapolative to other family B members. However, φ29 DNA polymerase is endowed with two distinctive features: high processivity and strand displacement capacity

* Correspondence should be addressed to mdevega@cbm.uam.es and msalas@cbm.uam.es

that allow it to replicate the viral genome from a single binding event, without requiring the assistance of unwinding and processivity factors. Recent crystallographic resolution of the structure of the apo and binary/ternary complexes of ϕ29 DNA polymerase, together with the biochemical studies of site-directed mutants, have given insights into the structural basis responsible for the coordination of the processive polymerization and strand displacement. In addition, such structures have provided the mechanism of translocation of family B DNA polymerases. Another difference with respect to the rest of replicases is the ability of ϕ29 DNA polymerase to use a protein (terminal protein, TP) as primer, circumventing the end replication problem. Recent resolution of the structure of the ϕ29 DNA polymerase/TP heterodimer, together with the biochemical analysis of chimerical DNA polymerases and TPs, have given the clues of the specificity of the interaction between both proteins, suggesting a model for the transition from initiation to elongation. We will also discuss how the basic research on the ϕ29 DNA polymerase properties have led to the development of DNA amplification technologies based on this outstanding enzyme.

In: Genetics Research Summaries
Editors: L. E. Hewitt and J. W. O'Neil
ISBN: 978-1-62257-646-3

Chapter 26

Bacterial Replicative DNA Helicases

***Subhasis B. Biswas*[1], *Jessica J. Clark*[1], *Deepa S. Kurpad*[2] *and Esther E. Biswas*[2]**
[1]Department of Molecular Biology, Graduate School of Biomedical Sciences, University of Medicine and Dentistry of New Jersey, Stratford, NJ, US
[2]Program in Biotechnology, Department of Bioscience Technologies Thomas Jefferson University, Philadelphia, PA, US

Research Summary

Replicative DNA helicases are energy-transducing enzymes, which act as the engine of the cellular replication apparatus and unwind duplex DNA in the replication fork. These enzymes are true multifunctional enzymes that are responsible for organization and execution of DNA replication from initiation to termination. The most well studied of them all is DnaB helicase of *Escherichia coli*, which is the prototype of this class of DNA helicases. Early genetic studies of the *dnaB* gene defined important roles of DnaB protein in the elongation stage of DNA replication. DnaB protein acts as the organizer of the replisome by engaging in multiple protein-protein and protein-DNA interactions during the initiation, elongation, and termination stages of DNA replication. DnaB protein and its orthologs have homo-hexameric structure in solution and in the ssDNA-bound state. This homo-hexameric subunit structure allows for the formation of a ring-structure; ssDNA passes through the ring's central hole. Formation of this unique protein-ssDNA complex is a key element in this enzyme's ability to carry out its unique functions. Due to its stable ring structure, *E. coli* DnaB helicase requires the assistance of DnaC protein in order to form the initial DnaB•ssDNA complex. It forms a DnaB•DnaC•ATP complex that attenuates its nucleotidase activity but helps in binding ssDNA. DnaC protein is released upon slow ATP hydrolysis leading to organization of the replication proteins in the fork and replisome formation. DnaB protein engineers the assembly of the replisome through its specialized interaction with DnaG primase and DNA polymerase III holoenzyme. DnaB•ssDNA complex formation stimulates its ATPase activity of DnaB hexamer several fold. The energy of ATP hydrolysis powers translocation of DnaB protein on ssDNA in a 5′→3′ direction and helps unwind duplex DNA during translocation. The rapid and processive DNA unwinding also requires direct participation of a topoisomerase to remove the DnaB-generated super-twists without which the DNA unwinding and the replication fork would likely stall. In this review article, we summarize the structural and functional properties of the *E. coli* replicative DNA helicase and compare it to analogous replicative helicases from other gram-negative and gram-positive bacteria.

In: Genetics Research Summaries
Editors: L. E. Hewitt and J. W. O'Neil
ISBN: 978-1-62257-646-3

Chapter 27

Effect of Substrate Traps on Hepatitis C Virus NS3 Helicase Catalyzed DNA Unwinding: Evidence for Enzyme Catalyzed Strand Exchange

Ryan S. Rypma, Angela M. I. Lam and David N. Frick*
Department of Biochemistry and Molecular Biology
New York Medical College, Valhalla, NY, US

Research Summary

The helicase encoded by the hepatitis C virus (HCV) is shown in this chapter to catalyze homologous DNA strand exchange. Single-stranded DNA oligonucleotides complementary to either the short or long strand of a partially duplex DNA substrate affected the rate and extent of unwinding catalyzed by either the full-length HCV NS3 protein fused to a portion of HCV NS4A, or a truncated NS3 protein lacking the protease domain. The oligonucleotides did not, however, sequester HCV helicase and prevent it from separating the original duplex after a single binding cycle. Furthermore, when DNA oligonucleotides were pre-incubated with HCV helicase, they did not prevent subsequent duplex separation, indicating that the enzyme catalyzes strand exchange. The protease portion of NS3 was not needed for this strand exchange. Fluorescent DNA substrates were further used to directly monitor both this ssDNA assisted unwinding and homologous strand exchange, and the effect of a protein trap (poly(U) RNA) on HCV helicase-catalyzed strand exchange was examined. The results demonstrate that HCV helicase can simultaneously bind at least three DNA strands and imply that HCV NS3 helicase could play an important role not only in viral RNA unwinding, but also in the folding of the HCV genome.

* To whom correspondence should be addressed: Department of Biochemistry and Molecular Biology, New York Medical College, Valhalla, NY 10595. Tel.:914-594-4190; Fax: 914-594-4058; E-mail: David_Frick@NYMC.edu

In: Genetics Research Summaries
Editors: L. E. Hewitt and J. W. O'Neil
ISBN: 978-1-62257-646-3

Chapter 28

Discovery of Orphan Helicases and Deorphanization by Genome-Wide Analyses in Two Model Organisms, *S. Cerevisiae* and *C. Elegans*

***Toshihiko Eki**,[1] *and Fumio Hanaoka*[2]**
[1]Division of Life Science and Biotechnology, Department of Ecological Engineering, Toyohashi University of Technology, Toyohashi, Aichi, Japan
[2]Department of Life Science, Faculty of Science, Gakushuin University, Mejiro, Toshima-ku, Tokyo, Japan

Research Summary

Eukaryotic DNA helicases are members of the helicase superfamily together with two other functional classes, RNA helicases and chromatin remodeling ATPases. Members of this superfamily play crucial roles in various nucleic acid- and chromatin-mediated cellular processes such as DNA replication, repair and recombination, pre-mRNA splicing, ribosome biogenesis, RNA interference, and chromatin remodeling. Since these reactions are essential for the maintenance, expression, and regulation of genetic information in the chromosome, dysfunctional helicase genes may lead to genetic diseases such as cancer. Indeed, genetic mutations of the human RecQ-like BLM helicase and WRN DNA helicase result in Bloom syndrome and Werner syndrome, which are characterized by the early development of various cancers and premature aging, respectively. Most helicases share conserved amino acid sequence motifs; their corresponding genes are classified into five superfamilies (SF1-SF5) based on the occurrence and characteristics of conserved motifs.

The genome sequences of two representative eukaryotes, the budding yeast *Saccharomyces cerevisiae* and the nematode *Caenorhabditis elegans* were determined in

* Tel: 81-532-44-6907; Fax: 81-532-44-6929; E-mail: eki@eco.tut.ac.jp

1996 and 1998, respectively. Subsequently, high-throughput genome-wide analyses have been systematically performed to clarify the cellular roles of 6000 and 19000 genes discovered in each genome, including gene expression profiling, proteome analyses, comprehensive analyses of loss-of-function phenotypes, and genetic interaction analyses. A huge amount of functional data from these studies has been analyzed and deposited in public databases such as the *Saccharomyces* Genome Database (SGD) and the WormBase. However, a few decades after the initial sequencing, a large number of genes remain functionally-unknown (i.e., orphan) in both organisms; for instance, over 1000 genes are uncharacterized even in yeast. A large number of helicase-like genes have been found in the sequenced eukaryotic genomes, including many genes encoding orphan helicase-related proteins. Because of the biological importance of the helicase family, a comprehensive analysis of the functions of helicase family members in two model organisms, *S. cerevisiae* and *C. elegans*, has been performed. In this Short Communication, we briefly summarize the results from studies of helicase-like proteins in both organisms and describe the effectiveness and limitations of a genome-wide analysis of orphan family members.

In: Genetics Research Summaries
Editors: L. E. Hewitt and J. W. O'Neil
ISBN: 978-1-62257-646-3

Chapter 29

Gene Regulation and Early Developmental Gene Expression in Vertebrate

Hongshi Yu and Shuliang Cui
Department of Zoology, The University of Melbourne
Royal Parade, Parkville, Victoria, Australia

Research Summary

Developmental gene regulation is to elucidate the mechanisms of spatio-temporal gene expression in organisms during development and disease occurring. This chapter is focusing on the regulation of early developmental gene expression based on the latest progress of vertebrate developmental studies. The fate of germ cells in extra-embryonic ectoderm is determined during PGC formation by predetermined germ plasm in the oocyte, from which VASA, the DEAD box family protein of ATP-dependent RNA helicase is identified as a regulator in germline cell specification, spermatogenesis, RNA splicing and post-translational degradation, and cell growth. The regulation of germline cell growth needs the multi-functional growth factor LIF, maintaining the pluripotency of ES cells in vitro and up-regulating its expression during implantation suggested the involvement of LIF in the event, which was further supported by direct evidence from gene knockout. The gonad in the early fetal life as one tissue, indifferent and indistinguishable by morphology, has two fates, making it a unique regulatory model of gene expression. Gene regulation and their interactions of many genes involved in this process including SRY, SOX9, WT1, FGF9, WNT4, DAX1 and DHH, and their regulating roles and interaction during sexual development will be discussed, particularly regulatory roles in alternative splicing and signal transduction pathway in gonadal development.

In: Genetics Research Summaries
Editors: L. E. Hewitt and J. W. O'Neil

ISBN: 978-1-62257-646-3

Chapter 30

Regulation of Gene Expression during Aging

Eugenia Villa-Cuesta *
Department of Ecology and Evolutionary Biology,
Brown University, Providence, RI, US

Research Summary

Genetic control of proliferation, morphogenesis, and differentiation during development of multicellular organisms is crucial for the proper formation of adults. Regulation of developmental gene expression, however, goes beyond the developmental stages of an organism. The rate at which organisms age for example, is also regulated by many developmental genes.

While it is intuitive to consider aging a simple byproduct of accumulated wear and tear on the organism, it nevertheless has been shown that altering expression of single genes can extend life span significantly; demonstrating that the aging process can be genetically influenced. Many genes involved in developmental processes modulate, later in life, adult aging. For instance, altered expression of genes affecting endocrine signaling, stress responses, metabolism, and growth during developmental stages can increase the life span of model organisms. Furthermore, the study of these genes has revealed evolutionarily conserved pathways for the modulation of aging. In contrast to the precise genetic regulation that occurs during development, life span, while genetically regulated, is not so tightly controlled. For example, there are significant differences in aging phenotype even in monozygotic human twins. In this chapter, I will discuss whether aging is mainly due to an organism's post-embryonic developmental process or a haphazard process, and I will describe and compare the regulation of genetic pathways involved in both development and aging. In doing so, I will consider what these two fields of science can learn from each other to progress our understanding of the regulation of these genes.

* Eugenia Villa-Cuesta@brown.edu

In: Genetics Research Summaries
Editors: L. E. Hewitt and J. W. O'Neil

ISBN: 978-1-62257-646-3

Chapter 31

Developmental Regulation of Sensory Receptor Gene Expression

Simon G. Sprecher
Department of Biology, Center for Developmental Genetics, New York University, New York, NY, US

Research Summary

Establishment of specific cell fates requires orchestrated interaction of an array of transcription factors and signaling pathways which incorporate major developmental roles. The transition of proliferating precursor or immature cells into a certain cell type and terminal differentiation has been studied in great detail on various neuronal cell types, due to the large degree of diversity, complex functional roles, and intrinsic properties of cells in the nervous system. In the peripheral nervous system sensory specificity is established by the choice of an immature, but committed, postmitotic progenitor to express a specific sensory receptor gene. Findings of sensory receptor gene regulation stemming form the olfactory system and visual system and in mouse and fruit fly provide insight in how this highly complex process is regulated and genetically controlled. After the initial decision to become a sensory neuron the cell then decides which receptor gene to express and subsequently maintain the expression of this given receptor gene. Genetic mechanisms for this choice depend upon transcriptional regulators which are expressed in subtype of sensory neurons, thus may provide a combinatorial code to orchestrate the expression of a specific receptor gene: For instance in the fly visual system, where an array of six *rhodopsin* genes can be expressed a combinatorial code of transcription factors is required for the sensory receptor gene regulation. Interestingly during larval and adult stages the regulation of *rhodopsins* makes use of distinct developmental genetic program. Regulatory regions of *rhodopsins* display a bipartite architecture with a proximal domain required for general PR expression and a distal domain encoding subtype specificity. In the human retina cone cells can express L and M opsin genes, which are located in close proximity on the chromosome. Regulation of L and M opsin depends upon "locus control regions" (LCR) a common long range *cis*-acting element regulating both genes. Interestingly, in the mouse retina *Opsins* are not clustered and co-expression of two *Opsins* genes occurs. In the mouse olfactory system, an array of over 1200 Odorant receptor (OR) genes can be expressed. OR genes are often arranged in clusters along the chromosome and seem to depend on distant and local *cis*-acting elements. Interestingly only one of the two parental copies of an OR gene is expressed, resulting in monoallelic expression of the gene. Taken together the choice of a sensory neuron to adopt a specific sensory specificity depends upon the complex interaction of *cis*- and *trans* acting factors. Even though generally only one form of sensory receptor gene is expressed, various mechanisms may be acting to achieve a similar outcome of sensory receptor gene regulation. Moreover recent findings reveal that sensory receptor genes can be co-expressed and that this co-expression is genetically controlled, thus adding an additional layer of complexity in the regulatory properties of sensory receptor genes.

In: Genetics Research Summaries
Editors: L. E. Hewitt and J. W. O'Neil
ISBN: 978-1-62257-646-3

Chapter 32

Normal and Injury-Induced Gene Expression in the Developing Postnatal Rat Inner Ear

Johann Gross [1], ***Ralf-Jürgen Kuban*** [2], ***Ute Ungethüm*** [2] ***and Birgit Mazurek*** [1]
[1]Department Othorhinolaryngology, Charité-Universitätsmedizin Berlin, Molecular Biology Research Laboratory
[2]Charité-Laboratory of Functional Genomics, Berlin, Germany

Research Summary

This chapter deals with an organotypic culture system to examine transcriptional events contributing to cell survival of the organ of Corti (OC), the modiolus (MOD) and the stria vascularis (SV) of newborn rats. mRNA profiling using Affymetrix gene chips was carried out in tissue obtained immediately after preparation and after 24 h in culture. The probe sets of 45 genes were subjected to a cluster analysis. A number of identified genes represented three major processes associated with the preparation of the cultures: mechanical injury and inflammation, hypoxia and excitotoxicity. The inflammatory response ontology was represented by inflammatory cytokines Interleukin-1beta (Il-1b), Interleukin 6 (Il-6), TNF-alpha converting enzyme (Tace) and Intercellular adhesion molecule (Icam). The hypoxia response was represented by the increase of Hypoxia-inducible factor-1 alpha (Hif-1a),

Glucose transporter 1 (Glut1) and Glucose transporter 3 (Glut3). The excitotoxic damaging process included changes in the glutamate transporters and NMDAR receptors. We identified Tace expression as a novel gene in the inner ear with a potentially important role in inner ear injury. The MOD region belongs to the most vulnerable regions of the ear characterized by a particularly high increase of Il-1b, Il-6 and Hif-1alpha mRNA expression.

Cell survival in culture is maintained by a complex regulation of pro-death genes on the one hand and protective genes on the other. In general, genes encoding proteins involved in triggering or executing cell death are downregulated and genes encoding protective acting proteins are upregulated. We found caspase 2, caspase 6 and calpain downregulated and the mitochondrial superoxide dismutase Sod2, the heat shock proteins Hsp27 and Hsp 70 and the insulin like growth factor binding proteins Igfbp3 and Igfbp5 upregulated. For two genes (Tace, Bax) we observed a differential response of coding and non-coding sequences. These data provide new insights into the role of the various members of the pro-death and pro-survival genes in protecting inner ear cells from injury-induced damage during the developing period.

In: Genetics Research Summaries
Editors: L. E. Hewitt and J. W. O'Neil

ISBN: 978-1-62257-646-3

Chapter 33

The TGF-B Superfamily: A Multitask Signalling Pathway for the Animal Kingdom

Marco Patruno[*]
Department of Experimental Veterinary Sciences, Faculty of Veterinary Medicine, Legnaro, PD, Italy

Research Summary

The TGF-β superfamily consists of numerous members, including TGF-β proper, bone morphogenetic proteins (BMP) and growth differentiation factors (GDF). All TGF-β are dimeric cytokines present a biological active carboxy terminal domain of 110–140 amino acids following proteolysis.

Bone morphogenetic proteins (BMPs), first identified for their involvement in vertebrate bone formation, are now widely recognized as key factors in the regulation of many fundamental developmental processes in all deuterostomes. The active gradient established by BMP secreted ligands is one of the essential factors responsible for generating the positional information that underlies developmental patterning, including the regeneration of lost parts.

Myostatin or GDF-8, a recently discovered GDF subfamily member, acts as a negative regulator in maintaining the mammalian proper muscle mass during both embryogenesis and post-natal muscle development. Unlike most other members of the BMP/GDF superfamily, mammalian myostatin is secreted as a latent complex, usually linked to regulatory proteins, and its mature dimer produces an effect almost exclusively on muscle tissue.

The present chapter deals with the importance of the TGF-β family of growth factors in relation to regulatory spheres through the animal kingdom, focusing in particular on the expression of BMP molecules during regeneration and myostatin in "non-canonical" animal models; it also focuses on the regulative actions of myostatin during the development of

[*] E-mail: marco.pat@unipd.it

vertebrates and in different experimental conditions, including *in vitro* chick co-culture and endurance training. Data available in literature indicate that there is substantial scope for future research in the area of TGF-β /myostatin linked to development.

"There is grandeur in this view of life, with its several powers, having been originally breathed by the Creator into a few forms or into one; and that, whilst this planet has gone cycling on according to the fixed law of gravity, from so simple a beginning endless forms most beautiful and most wonderful have been, and are being evolved." (*Charles Darwin, The Origin of Species*).

In: Genetics Research Summaries
Editors: L. E. Hewitt and J. W. O'Neil
ISBN: 978-1-62257-646-3

Chapter 34

Alpha-Foetoprotein: It's all About Timing

***De Mees Christelle*[1] *and Streel Emmanuel*[2]**
[1]Université Libre de Bruxelles, Institut de Biologie et Médecine Moléculaires, Gosselies, Belgium
[2]Université Libre de Bruxelles, ISM, Campus Erasme , Brussels, Belgium

Research Summary

Alpha-foetoprotein (AFP) is a well known diagnostic biomarker used in medicine to detect foetal developmental anomalies such as neural tube defects or Down's syndrome, or to follow the development of tumors such as hepatocellular carcinomas. However, the role of AFP goes way further than that. AFP is involved at least in rodents in the correct differentiation of the female brain, through its estrogen binding capacity. This chapter present an overview of what is known about the regulation of the *Afp* gene, describes the phenotype of the AFP knock-out (AFP KO) mouse and offers an overview of other mouse models available to study estrogen function.

Being in the right place, at the right time, is a key factor for alpha-foetoprotein (AFP). Firstly, because it is involved in major events occurring during narrow time-windows, such as sexual differentiation of the female brain. Secondly, because AFP is expressed in an onco-foetal way.

In: Genetics Research Summaries
Editors: L. E. Hewitt and J. W. O'Neil

ISBN: 978-1-62257-646-3

Chapter 35

Molecular Mechanisms of Cell-Specific Expression of Neuropeptide Hormone Genes, *DH-PBAN* and *PTTH* in the Silkworm, *Bombyx Mori*

Kunihiro Shiomi
Division of Applied Biology, Faculty of Textile Science and Technology, Shinshu University, Ueda, Nagano, Japan

Research Summary

Neurosecretory cells play critical roles in different and specific physiological and behavioral processes via spatiotemporal regulation of neurohormone secretion. In insects, diapause and metamorphosis are induced by neuropeptide hormones secreted from a few neurosecretory cells that project intrinsic axons to intrinsic neurohemal sites. Diapause hormone (DH) is responsible for induction of embryonic diapause in *Bombyx mori*. The diapause hormone-pheromone biosynthesis activating neuropeptide gene, *DH-PBAN,* is expressed exclusively in seven pairs of DH-PBAN-producing neurosecretory cells (DHPCs) on the terminally differentiated processes of the subesophageal ganglion (SG). On the other hand, prothoracicotropic hormone (PTTH) plays a central role in controlling molting and metamorphosis in *Bombyx mori* by stimulating the prothoracic glands to synthesize and release the molting hormone, ecdysone. The *PTTH* gene is constantly expressed during larval-pupal development, and the peptide is produced exclusively in two pairs of lateral PTTH-producing neurosecretory cells (PTPCs) in the brain. To help reveal the regulatory mechanisms of cell-specific expression of *DH-PBAN* and *PTTH*, we identified *cis*-regulatory elements that regulate expression in DHPCs and PTPCs, respectively, using a recombinant baculovirus (AcNPV)-mediated gene transfer system and a gel-mobility shift assay. Interestingly, *Bombyx mori* Pitx (BmPitx), a bicoid-like homeobox transcription factor, binds the 5'-upstream sequence of both *DH-PBAN* and *PTTH* and activates gene expression. This article describes the regulatory mechanisms of cell-specific expression of the neuropeptide hormone genes involved in diapause and metamorphosis in *Bombyx*.

In: Genetics Research Summaries
Editors: L. E. Hewitt and J. W. O'Neil

ISBN: 978-1-62257-646-3

Chapter 36

Gene Expression Analysis during Development by High-Throughput Methods

***Francesca Amati*[1], *Giovanni Chillemi*[2] and *Giuseppe Novelli*[1]**
[1]Department of Biopathology and Diagnostic Imaging,
Tor Vergata University, Rome, Italy
[2]CASPUR, Rome, Italy

Research Summary

Regulation of gene expression during embryogenesis and development is a crucial clue for a normal anatomy and physiology. In fact, very little is known regarding factors that influence and regulate developmental gene expression. Similarly, there is little information available concerning the effects of a coordinate expression of a group of functionally related genes.

The analysis of temporal patterns of gene expression in embryos is essential for the understanding of the molecular mechanisms that control development. This scientific field has been innovated by the combined use of experimental high-throughput methods, such as DNA microarrays, and bioinformatic methods that take advantage of the completion of the human genome sequence, along with the genomes of related species. Microarray analysis, in fact, provides a large amount of data -at molecular level- that once acquired, must be functionally integrated in order to find common patterns within a defined group of biological samples. Following the enormous number of data obtained from these experiments, a new type of comparative embryology is now emerging, and it is based on the comparison of gene expression patterns. The sequencing of several new genomes, the increasing computational power and new bioinformatic algorithms cooperate to overcome some of the intrinsic difficulties in the study of gene regulation, thus permitting, for example, to identify regulation sites located far away from the genes. Recent bioinformatic methods applied to gene regulation are reviewed that either follow the "single species, many genes" approach or the "single gene, many species" one.

In this chapter we would review the new application of DNA microarray and bioinformatics to define a new combinatorial approach for analysis of gene expression during development.

In: Genetics Research Summaries
Editors: L. E. Hewitt and J. W. O'Neil

ISBN: 978-1-62257-646-3

Chapter 37

Gene Expression Regulation in the Developing Brain

***Ching-Lin Tsai*[1] *and Li-Hsueh Wang*[2]**
[1]Department of Marine Biotechnology and Resources,
National Sun Yat-sen University, Kaohsiung, Taiwan
[2]National Museum of Marine Biology and Aquarium,
Checheng, Pingtung, Taiwan

Research Summary

The developing central neural circuits are genetically controlled and initiated by developmental signals. Recent progress in molecular and cellular developmental biology provides evidence of how the brain is feminized or masculinized during the critical developmental period. Research into the development of brain architecture requires experimenting with animals, specifically, interfering with normal development and with environmental conditions. Drosophilae, sea urchins, and metazoans are simple invertebrates used for standard research models. Recently, the teleosts, bony fish with biological and genomic complexity found in the higher vertebrates, have become important models for developmental and molecular neurobiology studies. As in mammals, sexual dimorphic genetic expression is found in the developing brain of teleosts. The cellular and synaptic organization of brain architecture is determined by the genomic program and triggered by environmental cues such as photoperiod and temperature. This review highlights some of the methodological issues related to current findings about the gene expression regulation involved in the complex process of neural development, particularly in brain-sex differentiation.

In: Genetics Research Summaries
Editors: L. E. Hewitt and J. W. O'Neil
ISBN: 978-1-62257-646-3

Chapter 38

Expression and Action of SRY during Gonadal Sex Differentiation in the Mouse

Teruko Taketo and Chung-Hae Lee
Departments of Surgery and Biology, McGill University,
Royal Victoria Hospital, Montreal, Quebec

Research Summary

SRY/Sry, a single-copy gene on the Y-chromosome, was identified to play the critical role in initiating testicular differentiation during gonadal development in humans and mice two decades ago. Nonetheless, neither the regulation of *Sry* expression nor the mode of SRY action during gonadal differentiation is well understood. The B6.Y^{TIR} mouse carries a Y-chromosome originally from a *Mus musculus domesticus* mouse caught in Tirano, Italy (Y^{TIR}) and the X-chromosome and autosomes from the C57BL/6J (B6) inbred mouse strain, which belongs to *Mus musculus molossinus*. It has been demonstrated that the SRY protein is expressed normally both in pattern and onset, yet, B6.Y^{TIR} mice develop only ovaries or ovotestes. Therefore, this mouse model provides an opportunity to study the mechanism of SRY action during gonadal sex determination. We hypothesize that the testis determining pathway in the B6.Y^{TIR} gonad is impaired by at least two mechanisms that act synergistically. First, *Sry* transcript levels from the Y^{TIR}-chromosome are reduced on the B6 genetic background. Second, polymorphisms of *Sry* sequences lead to inefficient biological activity of the SRY protein encoded on the Y^{TIR}-chromosome. Both dysfunctions are requisite to impairing testicular differentiation.

In: Genetics Research Summaries
Editors: L. E. Hewitt and J. W. O'Neil

ISBN: 978-1-62257-646-3

Chapter 39

The Fibroblast Growth Factor (FGF) Gene Families of Japanese Medaka (*Oryzias latipes*)

Asok K. Dasmahapatra and Ikhlas A. Khan
National Center for Natural Product Research
Department of Pharmacology, School of Pharmacy
University of Mississippi, Oxford, MS, US

Research Summary

Fibroblast growth factors (FGFs) constitute a large family of signaling polypeptides that play critical roles in development. During morphogenesis, FGFs are involved in cell proliferation, differentiation and migration; however, in adults these proteins function as homeostatic factors. FGFs mediate their functions through a cell surface receptor, the fibroblast growth factor receptors (FGFRs), which are a member of the tyrosine kinase superfamily. Both the *FGF* and *FGFR* gene families are identified in multicellular organisms but not in unicellular ones and have expanded greatly during evolution. FGF gene organization is highly conserved among vertebrates. In human and mouse, the FGF gene family consists of 22 members; however, in zebrafish (*Danio rerio*) there are 27 identified *fgf* members. Japanese medaka *(Oryzias latipes),* like zebrafish, is a small aquarium fish used as a model organism in vertebrate development. During evolution, these two fish species (zebrafish and Japanese medaka) were separated from their last common ancestor about 110 million years ago. The medaka genome is only half (800 Mb) of the zebrafish genome (1700 Mb). We have searched medaka genome data bases and identified 28 *fgf* genes in this species of which nine are paralogs. We have done a phylogenetic and conserved gene location (synteny) analysis of the identified *fgf* genes of medaka and analyzed the evolutionary relationships of these genes with human *FGF* gene families.

In: Genetics Research Summaries
Editors: L. E. Hewitt and J. W. O'Neil

ISBN: 978-1-62257-646-3

Chapter 40

The Role of Internal Fluid Invironment for Regulation of Germ Genes Expression in Early Development of Some Cyprinid Fishes and their Intergeneric F1 Hybrids

A. M. Andreeva
Institute of Biology of Inland Waters,
Borok, Yaroslavl region, Russia

Research Summary

Connection between dynamics of yolk lipovitellin degradation and specific features of germ genes expression was studied in early development of intergenetic reciprocal F1 hybrids of the bream, roach and blue bream.

According to the modern view lipovitellin is the main protein of oocyte and embryo yolk. Its main function is reserve, nutritional and structural. But moreover lipovitellin is active component of internal fluid embryo invironment in wich embryo cells and germ genes are developing and expressing. There is a view that synchronous expression of parental alleles of genetic loci is related, as a rule, to kindred fish crossing, and asynchronous expression – to remote fish crossing. But when we analysed character of expression of some loci of intergeneric F1 hybrids (*6-Pgd*, *Ldh-B*, *β–Est-1, 2, 3*, *Aat-1*, *Me-1, 2* and others) with different expression time in embryogenesis, we show that character of loci expression is related to stage of development also.

As objects we used zygotes, embryos, larvas and frysof bream, roach, blue bream and intergeneric reciprocal F1 hybrids. Identification and analysis of enzymes activity and lipoviteelline properties were performed using methods of disk-, gradient and SDS-electrophoresis in polyacrylamide gel.

There are some arguments for regulation function of lipovitellin : first argument is connected with different expressions character (synchronous and asynchronous) of germ genes in early and late stages of embryogenesis; second - with different distribution of isoenzymes activity in early and late stages of embryogenesis; and third – with biocatalytic activity of lipovitelline because of oogenesis isoenzymes connected with lipovitelline by weak connection.

When first locus expression was timed to the early stages (blastodisk – gastrula), the gene parental alleles were activated asynchronously according to the maternal types. When the first expression was timed to later stages (yolk sac resorption), parenteral alleles were activated synchronously. In early development lipovitellin and oogenesis enzymes form the maternal (by origin) metabolic medium, which preferential activation of maternal alleles. At later developmental stages, when the yolk reserves are partially or fully resorbed and maternal proteins don' t play importance role, and germ proteins form new (germ) internal fluid invironment, in wich the embryonic genes are activated synchronously.

In: Genetics Research Summaries
Editors: L. E. Hewitt and J. W. O'Neil

ISBN: 978-1-62257-646-3

Chapter 41

Cloning and Expression Analysis of the Homeobox Gene *Abdominal-A* in the Isopod *Asellus Aquaticus*

Philipp Vick**, Axel Schweickert and Martin Blum
Institute of Zoology, University of Hohenheim,
Stuttgart, Germany

Research Summary

Regulation of embryonic axis patterning by Hox genes has been shown to be widely conserved among metazoans. In *Drosophila melanogaster* the Hox gene *abdominal-A* (*abd-*A) is important for the development of the legless abdomen. In contrast to the clear tagmata-correlated activity in insects the analysis of Hox expression patterns during crustacean development has turned out to be more diverse. While in the branchiopod brine shrimp *Artemia franciscana* the posterior genes show a more ancestral overlapping arrangement, this is not the case for the malacostracan isopod *Porcellio scaber*. In this more modern species *abd-A* is mainly restricted to the developing pleon. Here we present the cloning and expression pattern of the *abd-A* gene from the freshwater crustacean *Asellus aquaticus*. In contrast to the related isopod *Porcellio scaber*, *Asellus aquaticus* differs in the regulation of posterior segment patterning. While *Porcellio scaber* displays distinct and separate segments in the pleon, posterior segments are partially fused in *Asellus aquaticus* to yield a pleotelson. The *abd-A* signal was significantly reduced or absent in the pleotelson of *Asellus aquaticus*, while *abd-A* was expressed in the free segments of *Porcellio scaber*. An additional correlation between *abd-A* gene expression and patterning in these two species was found in that the orientation of walking legs was towards the posterior pole in segments expressing *abd-A*. *Asellus aquaticus* thus may provide a highly interesting and novel arthropod model organism to study evolution of segment identity and patterning.

---

* E-mail:pvick@uni-hohenheim.de

In: Genetics Research Summaries
Editors: L. E. Hewitt and J. W. O'Neil
ISBN: 978-1-62257-646-3

Chapter 42

Tidying up a Molecular Catastrophe: *In Vitro* and *In Vivo* Repair of Unique DNA-Protein Cross Links

David J. Baker, Arthur D. Riggs and Timothy R. O'Connor
Biology Department; Beckman Research Institute,
City of Hope Duarte, CA, US

Research Summary

Even small DNA-protein cross links (DPCs) can be more than 100 times the size of small base modifications. Most current strategies for the study of the elimination of DPCs focus on examination of massive numbers of protein adducts. Although DPCs represent a small fraction of total DNA lesions inflicted by damage such as radiation, these adducts are much more lethal to cells compared to most DNA modifications. Therefore, the elimination of DPCs is critical for cell survival, since the persistence of such lesions can lead to cell death mainly by posing a barrier to DNA and RNA synthesis. As DPCs are a minor overall component of most DNA damaging treatments, the study of their repair is complicated due to the low frequency at which they occur. To facilitate the investigation of DNA repair for this important DNA damage class, a few groups, including ours, have used a different tactic to evaluate the repair of DPCs in oligodeoxyribonucleotides at defined sites, which implicates nucleotide excision repair in this process. To extend the *in vitro* results and obtain *in vivo* information on cellular mechanisms that maintain genomic integrity, we have also constructed plasmids modified with a single DPC. We will evaluate the implications of these data with respect to therapies using DNA cross-linking agents. This review will summarize the results and understanding of DPC repair obtained from these studies.

In: Genetics Research Summaries
Editors: L. E. Hewitt and J. W. O'Neil
ISBN: 978-1-62257-646-3

Chapter 43

In Vivo Mutagenic Effects of Alkylating Agents Eliciting Different DNA-Adducts

Henriqueta Louro and Maria João Silva
Department of Genetics, National Institute of Health Dr. Ricardo Jorge, Lisbon, Portugal

Research Summary

Carcinogenesis is a complex, multistep process resulting from a series of mutations in critical genes controlling cell division, genome stability or cell death. It is well-established that most mutations result from interaction of chemical carcinogens with DNA to produce DNA adducts. Numerous genotoxicity assays have been conceived to measure the ability of a chemical to induce DNA lesions, mutations and chromosomal damage. Among these, *in vivo* studies of mutagenesis have been developed to account for the complexity of a whole-organism, providing valuable clues to the understanding of the mechanisms of mutation. The unique mutational spectrum produced by an agent in its target organ(s) is dictated mostly by the type and location of DNA adducts and by the repair framework triggered by the lesion, and thus can amplify the knowledge on the mechanisms of *in vivo* DNA repair, mutagenesis and carcinogenesis.

Using a *lacZ* plasmid-based transgenic mouse model, we have analyzed the frequency and spectrum of mutations induced by agents known to elicit different pre-mutagenic lesions in DNA. Additionally, we generated the *Parp1*$^{-/-}$ *lacZ* mouse model, with inactivation of the enzyme poly (ADP-ribose) polymerase (Parp1), and we tried to evaluate whether Parp1 deficiency influenced the *in vivo* mutagenic effect of the monofunctional alkylating agent N-Methyl-N-Nitrosurea (MNU).

Based on those studies we provided evidence that cisplatin, an anticancer drug, has mutagenic activity *in vivo* and produces a specific profile of base substitutions and small deletions/insertions in the liver of *lacZ* mice, in agreement with its preferential DNA binding sites. In a subsequent study, we showed that MNU increased the mutant frequency in *Parp1*$^{+/+}$ and *Parp1*$^{-/-}$ *lacZ* mice. Interestingly, our data evidenced a shift in the pattern of MNU-induced mutations in testis of *Parp1*$^{-/-}$ as compared to *Parp1*$^{+/+}$ *lacZ* mice, consisting of a higher frequency of deletions/insertions and a lower frequency of point mutations. In contrast, the MNU-induced point mutation spectra in testis were independent of the Parp1 status.

Those studies illustrate how the data from *in vivo* transgenic assays can help to clarify specific pre-mutagenic adducts, DNA repair functions and mutational events that may be involved in the mutagenicity of human carcinogens.

In: Genetics Research Summaries
Editors: L. E. Hewitt and J. W. O'Neil
ISBN: 978-1-62257-646-3

Chapter 44

DNA Adduct Formation and Genotoxicity Assessment in Plants

Frédérique Weber-Lotfi, Pierre Guillemaut, Benoît Rether, and André Dietrich*
Institut de Biologie Moléculaire des Plantes,
CNRS/Université de Strasbourg, Strasbourg, France

Research Summary

Numerous physical and chemical stress factors, endogenous or exogenous, challenge living organisms. Cells take up various xenobiotic compounds which are subsequently detoxified through metabolization, conjugation and elimination. However, metabolization processes generate highly reactive intermediates which can covalently bind to the DNA, resulting in bulky addition products called "adducts". In animal cells, when they escape elimination by the DNA repair systems, adducts perturb gene expression and induce mutations. Formation of DNA adducts has therefore been highly correlated with carcinogenesis. Originally considered only in the animal kingdom, these genotoxic processes also occur in plants and are likely to impair their gene expression. The literature in that domain is still limited, but DNA adduct formation appears to be a general response of plants to organic chemical exposure, whether in controlled conditions or in the field. In particular, common carcinogenic pollutants and pesticides promote the formation of DNA adducts in plants. The modified nucleotides generated from a given chemical can be similar or different between plant and animal cells, reflecting similarities or differences in the metabolization pathways of the exogenous compounds. On the other hand, it seems possible to identify the origin of some DNA adducts occurring in field plants by co-migration with samples from plants treated with a defined chemical in controlled conditions. Experimental data indicate

* Corresponding author: Institut de Biologie Moléculaire des Plantes, CNRS/Université de Strasbourg, 12 rue du Général Zimmer, F-67084 Strasbourg, France. phone ++33 3 88 41 72 41, fax ++33 3 88 61 44 42. e-mail andre.dietrich@ibmp-ulp.u-strasbg.fr.

that the analysis of DNA adducts in test-plants can be a relevant and sensitive alternative to characterize oxidative and genotoxic stress from pesticides in crops or to evaluate the genotoxicity of the atmospheric environment. Cells can also eliminate DNA adducts thanks to DNA repair pathways. In human cells, the nucleotide excision repair (NER) pathway is known to eliminate bulky adducts. The NER pathway also exists in plants, but shows significant differences as compared to its mammalian counterpart.

In: Genetics Research Summaries
Editors: L. E. Hewitt and J. W. O'Neil

ISBN: 978-1-62257-646-3

Chapter 45

Diet, Life-Style Habits and Malondialdehyde–Deoxyguanosine Adducts in a Group of Subjects Resident in the Rayong Province, Thailand

M. Peluso[1,*], P. Srivatanakul[2], A. Munnia[1], M. Ceppi[3], A. Jedpiyawongse[2], S. Sangrajrang[2], S. Piro[1] and P. Boffetta[4]

[1]ISPO-Cancer Prevention and Research Institute, Florence, Italy
[2]National Cancer Institute, Bangkok, Thailand
[3]National Cancer Institute, Genoa, Italy
[4]International Agency for Research on Cancer, France

Research Summary

Most human cancers result from environmental exposures and gene-environment interactions. Diet is an important component of lifestyle and its role in the maintenance of good health and as a determinant of different kind of chronic diseases, including cancer, has been extensively analyzed. Diet can contribute to cancer risk through the consumption of food mutagens, including those generated by cooking. On the other hand, diet that emphasizes the consumption of whole grain foods, legumes, vegetables and fresh fruit and that limit animal fat has been associated with decreased cancer risk. Indeed, a general protective effect against cancers at different sites has been reported for a number of fresh fruit and vegetables. In

* Corresponding author: Marco Peluso, Cancer Risk Factor Branch, Analitycal Citology and Biomolecular, ISPO - Cancer Prevention and Research Institute, Via Cosimo il Vecchio 2, 50139 Florence, Italy. Tel.: + 39 0 55 32 69 78 67; fax: + 39 0 55 32 69 78 79. E-mail address:m.peluso@ispo.toscana.it.

particular, fresh fruit is a food rich of phytochemicals that can act against reactive oxygen species (ROS) formation and induce the expression of DNA repair systems.

Among life-style habits, tobacco cigarette smoking is an established cancer risk factor. In addition to be a relevant source of mutagens, tobacco smoke is also a major source of oxidative stress and ROS, capable to cause oxidative DNA damage. ROS can also induce lipid peroxidation (LPO), whose by-products can lead to endogenous DNA damage formation.

MDA is a natural product of LPO, also formed during prostaglandins (PGH_2) biosynthesis via cyclooxygenase (COX-1 and COX-2). MDA is an aldehyde capable to interact with cellular constituents including DNA, inducing the formation of M_1dG adducts. The reaction with deoxyguanosine (dG) origins the formation of 3-(2-deoxy-β-D-erythro-pentafuranosyl)pyrimido[1,2-α]purin-10(3H)-one, e.g. the M_1dG adducts. In addition to LPO and PGH_2 biosynthesis, the formation of M_1dG adducts can be induced by ROS through the production of base propenal intermediates.

M_1dG adducts have been shown to block the replication of DNA and induce base pair and frameshift mutations. Moreover, increased levels of M_1dG adducts have been found in number of cancer patients. Higher amount of M_1dG adducts have been also associated to worst surviving in lung cancer patients. Thus, the analysis of M_1dG adduct formation is considered a promising biomarker that can reflect the whole oxidant status of an organism.

In the present study, we have investigated the relationship between tobacco smoking, the consumption of fresh fruit and the formation of M_1dG adducts in a group of people resident in Map Ta Phut (MTP), Rayong province, Thailand, including 53 residents in the Map Ta Phut Industrial Estate (MIE) area, and 46 individuals living in a control district of the same province without industrial exposures. The DNA adduct analysis was performed using the ^{32}P-DNA postlabelling assay (Munnia et al., 2006). The potential involvement of the polymorphism of glutathione S-transferase Mu-1 (GSTM1) in the formation of endogenous DNA adducts has been also evaluated.

Our findings indicate a statistically significant effect of tobacco smoke on M_1dG formation ($p < 0.05$). The increased intake of fresh fruit was inversely associated to the M_1dG adduct production ($p < 0.05$). Then, the GSTM1 polymorphism was not associated with the formation of M_1dG adducts.

Our data indicate that tobacco smoking habits can induce an enhanced formation of M_1dG adducts, possibly through the production of ROS and LPO. Conversely, the consumption of fresh fruits was associated to a decreased formation of DNA damage. The increased formation of M_1dG adducts in current smokers can be related to an excess of LPO induced from tobacco smoke constituents, as well from ROS produced during carcinogen metabolism. The protective effects of fresh fruit on M_1dG adduct production may act on different biological mechanisms, for example by the induction of enzymes involved in DNA repair and by antioxidant effects. The GSTM1 polymorphism could be not relevant for the frequency of M_1dG adducts.

In: Genetics Research Summaries
Editors: L. E. Hewitt and J. W. O'Neil
ISBN: 978-1-62257-646-3

Chapter 46

Drug-DNA Adducts Formed by Formaldehyde Activation of Anthracyclines and Related Anti-Cancer Agents

Don R. Phillips[1], *Lonnie Swift*[1,2], *Kate Coldwell*[1,3], *Damian Spencer*[1], *Rebecca Bilardi*[1], *Benny Evison*[1], *Oula Mansour*[1], *Abraham Nudelman*[4], *Ada Rephaeli*[5], *Tad H. Koch*[6] *and Suzanne M. Cutts*[1]
[1]Department of Biochemistry, La Trobe University, Victoria, Australia
[2]Cancer Research UK Laboratories, Weatherall Institute of Molecular Medicine, University of Oxford, John Radcliffe Hospital, Oxford, UK
[3]Stem Cell Research, NHS Blood and Transplant Oxford, UK
[4]Chemistry Department, Bar Ilan University, Ramat Gan, Israel
[5]Felsenstein Medical Research Center, Sackler School of Medicine, Tel Aviv University, Beilinson Campus, Petach Tikva, Israel
[6]Department of Chemistry and Biochemistry, University of Colorado, Boulder, CO, US

Research Summary

The anthracyclines (doxorubicin, epirubicin, idarubicin and daunorubicin) are among the most widely-used anticancer chemotherapeutic agents. They are generally classified as topoisomerase II inhibitors but also have a variety of other targets in cells. It has been known for some years that the anthracyclines (particularly doxorubicin) are capable of forming DNA adducts, but the relevance and extent of these adducts in cells and their role in causing cell death has remained obscure. When the structure of adducts was solved, it became clear that formaldehyde was an absolute requirement for adduct formation. Adducts form primarily at GC sequences and involve an N-C-N aminal linkage between the drug and the N-2 of guanine, with the central carbon being derived from formaldehyde. The adducts have been extensively characterised *in vitro* by 2D-NMR, X-ray crystallography and mass spectrometry, and the aminal linkage has been shown to be reversible, with a half-life of up to 40 h. In cells, the DNA lesions produced appear to be dependent on two competing cellular pathways that can be switched from primarily topisomerase II lesions to primarily drug-DNA adducts by the use of formaldehyde-releasing prodrugs such as AN-9. The anthracyclines-DNA adducts are able to overcome the major forms of resistance to doxorubicin, are more cytotoxic than topoisomerase II lesions, and invoke conventional apoptotic pathways that are independent of topoisomerase II. Significant levels of adducts have now been detected in breast cancer cells (approximately 1,300 adducts per cell) following a 4 h treatment with clinically relevant levels of doxorubicin (25 nM) as a single agent, and this is elevated some 100-fold by co-treatment with formaldehyde-releasing prodrugs. The synthetic anthracendione anticancer agent, mitoxantrone, can also form adducts with DNA when activated by formaldehyde. These adducts exhibit specificity for CG and CA sequences but are rather unstable. More stable adducts have been detected with pixantrone, a second-generation mitoxantrone analogue that contains primary amino groups on the side-chain. However, the need to identify novel anthracenediones with enhanced anthracenedione-DNA adduct stability remains if these adducts ultimately are to be useful in the clinic.

In: Genetics Research Summaries
Editors: L. E. Hewitt and J. W. O'Neil
ISBN: 978-1-62257-646-3

Chapter 47

Differential Modulation of Benzo[*a*]pyrene-Derived DNA Adducts in MCF-7 Cells by Marine Compounds

***Jamal M. Arif*[1,*], *Mohammed Kunhi*[1], *Yunus M. Siddiqui*[1], *Khalid A. El Sayed*[2], *Khaled Y. Orabi*[3], *Amal Al-Hazzani*[1], *Mohammed N. Al-Ahdal*[1] and *Fahad M. Al-Khodairy*[1]**

[1]Biological and Medical Research, King Faisal Specialist Hospital and Research Institute, Riyadh, Kingdom of Saudi Arabia

[2]Department of Basic Pharmaceutical Sciences, School of Pharmacy, University of Louisiana at Monroe, Monroe, LA, US

[3]Department of Pharmaceutical Chemistry, Faculty of Pharmacy, Kuwait University, Safat, Kuwait

Research Summary

Despite several marine anticancer compounds to be DNA-interactive, causing from reductive DNA cleavage to DNA adduct formation, not much attention has been given to the DNA adduct as early biomarker in the screening and development of marine anticancer drugs. Most of these compounds have been tested *in vitro* by high-throughput cost-effective screening assays for their anticancer potential. In the present study, chemically diverse marine compounds were screened using benzo[*a*]pyrene (BP)-derived DNA adduct formation in MCF-7 cells. Briefly, MCF-7 cells were incubated with the marine compounds, namely manzamine A, sarcophine, aaptamine, verongiaquinol, curcuphenol, and curcudiol (10, 50

[*]To whom reprint requests should be addressed: Jamal M. Arif, PhD, MBC# 03-81, Biological and Medical Research, King Faisal Specialist Hospital and Research Center, P.O. Box 3354, Riyadh 11211, Kingdom of Saudi Arabia, Tel: 966-1-464-7272 ext. 32863; Fax: 966-1-442-7858, E-mail: jarif@kfshrc.edu.sa

and 100 μM) for 24 h followed by treatment with BP (0.5 μM). After 24 h re-incubation, cellular DNA was isolated and analyzed for BP-derived DNA adducts by ^{32}P-postlabeling technique. Interestingly, at 50 μM dose, only manzamine A, a probable antitumor compound, increased the BP-DNA adducts by 3-folds while curcuphenol and curcudiol lowered the BP-DNA adduction by up to 50%. Other compounds insignificantly modulated the BP-DNA adduction. At 10 μM, all the marine compounds were ineffective except aaptamine, which induced BP-DNA adduction by 2-fold. Further at 100 μM dose, all the compounds except manzamine A and verongiaquinol increased the BP-DNA adducts by up to 500%. In addition, verongiaquinol (50 μM) substantially down-regulated the levels of p53, p21, and p16, while manzamine A (50 μM), significantly inhibited the expression of p53. Aaptamine (50 μM) induced the p53 expression by 40%, however, other compounds were ineffective. All other marine compounds resulted in a differential response related to p21 and p16 expression. The residual DNA repair ability was almost completely abolished by manzamine A and verongiaquinol (50 μM) while other compounds seemed ineffective. Based on the preliminary results, it seems that these marine compounds *per se* may not be genotoxic but have potential of differentially modulating the carcinogen-derived DNA adducts enhancing the

In: Genetics Research Summaries
Editors: L. E. Hewitt and J. W. O'Neil

ISBN: 978-1-62257-646-3

Chapter 48

Potency of Air Pollutants at DNA Adduct Formation and Assessment by *In Vivo* Mutagenesis

Yasunobu Aoki, Akiko H. Hashimoto, Hiromi Sato and Michi Matsumoto
National Institute for Environmental Studies,
Research Center for Environmental Risk
Tsukuba, Ibaraki, Japan

Research Summary

Various mutagenic polycyclic aromatic hydrocarbons (PAHs) including nitro-PAHs, associated with suspended particulate matter (SPM), are possibly deposited in the lungs and other organs of residents in urban areas. To assess the mutagenicity of air pollution and its carcinogenic risk in the lung and other organs, it is necessary to assess the genotoxicity *in vivo* of the various PAHs and nitro-PAHs in urban air. In this chapter, we review the genotoxicity, such as DNA adduct formation, of air pollutants and its assessment by *in vivo* mutagenesis.

In: Genetics Research Summaries
Editors: L. E. Hewitt and J. W. O'Neil
ISBN: 978-1-62257-646-3

Chapter 49

GSTM1 and XRCC3 Polymorphisms: Effects on the Levels of Aflatoxin B1-DNA Adducts

Xi-Dai Long*[*1], Yun Ma[2], Jin-Gang Yao[3], Fu-Zhi Ban[4], Yuan-Feng Zhou[1], Zhong-Hui Lin[5], Hong-Dong Huang[6] and He-Ping Yang[7]

[1]Department of Pathology, Youjiang Medical College for Nationalities, Baise, Guangxi Zhuang Autonomous Region, China
[2]Department of Pathology, Guangxi Medical University, Nanning, Guangxi Zhuang Autonomous Region
[3]Department of Medicine, Affiliated Hospital of Youjiang Medical College for Nationalities, Baise, Guangxi Zhuang Autonomous Region, China
[4]Department of Medical Tests, Southwest Hospital of Youjiang Medical College for Nationalities, Baise, Guangxi Zhuang Autonomous Region, China
[5]Department of Medicine, Xiamen Zhong Hospital, Xiamen, Fujiang Province.
[6]Department of Medicine, the People's Hospital of Hunan Province, Changsha, China
[7]Department of Medicine, the Second Hospital of Central South University, Changsha, China

Research Summary

Aflatoxin B1 (AFB1), resulting in the formation of AFB1-DNA adducts, is a known human carcinogen. AFB1-exposed individuals with inherited susceptible carcinogen-metabolizing or repairing genotypes may experience an increased risk of genotoxicity. This study was designed to investigate whether the polymorphisms of two genes, namely, one

* Address correspondence to: Xi-Dai Long, Chengxiang Road No. 98, Baise City, Guangxi Zhuang Autonomous Region, China, 533000. Tel/Fax: +86 776 2849481. E-mail: xiaolonglong200166@yahoo.con.cn

metabolic gene Glutathione S-transferase M1 (GSTM1) and one DNA repair gene x-ray repair cross-complementing group 3 (XRCC3) codon 241, affected the levels of AFB1-DNA adducts in a Guangxi population (n = 966), from an AFB1-exposure area. AFB1-DNA adducts were measured by ELISA, and GSTM1 and XRCC3 codon 241 genotypes were identified by PCR-RFLP. The GSTM1-null genotype (adjusted odds ratio [OR] = 2.09; 95% confidence interval [CI] = 1.61–2.71) and XRCC3 genotypes with 241 Met alleles (namely, XRCC3-TM and –MM; adjusted ORs [95% CI] were 1.43 [1.08–1.89] and 2.42 [1.13–5.22], respectively) was significantly associated with higher levels of AFB1-DNA adducts. Compared with those individuals who did not express any putative risk genotypes as reference (OR = 1), individuals featuring all of the putative risk genotypes did experience significantly higher DNA-adduct levels (adjusted ORs were 2.87 for GSTM1-null and XRCC3-TM; 5.83 for GSTM1-null and XRCC3-MM). Additionally, there was a positive joint effect between XRCC3 genotypes and long-year AFB1 exposure in the formation of AFB1-DNA adducts. These results suggest that individuals with susceptible genotypes GSTM1-null, XRCC3-TM, or XRCC3-MM may experience an increased risk of DNA damage elicited by AFB1 exposure.

In: Genetics Research Summaries
Editors: L. E. Hewitt and J. W. O'Neil

ISBN: 978-1-62257-646-3

Chapter 50

Oxidative DNA Damage and DNA Adducts Induced by Polycyclic Aromatic Hydrocarbons and Aromatic Amines

Shosuke Kawanishi** and *Shiho Ohnishi
Faculty of Pharmaceutical Sciences, Suzuka University of Medical Science, Suzuka, Mie, Japan

Research Summary

Adduct formation has been considered a major factor of DNA damage by carcinogenic polycyclic aromatic hydrocarbons (PAHs) and aromatic amines with metabolic activation. Benzo[a]pyrene (BP), contained in tobacco smoke and air pollution, is one of the most investigated compounds as a carcinogenic PAH. It is considered to express genotoxicity by forming DNA adducts of its metabolite, BP-7,8-diol-9,10-epoxide (BPDE). In addition, BP may induce oxidative DNA damage. A metabolite of BP, benzo[a]pyrene-7,8-dione (BP-7,8-dione) causes oxidative DNA damage, including 8-oxo-7,8-dihydro-2'-deoxyguanosine (8-oxodG) formation.

Considered carcinogenic aromatic amines contained in cigarette smoke, 2-naphthylamine (2-NA) and 4-aminobiphenyl (4-ABP), have been known to form DNA adducts after metabolic activations. Nitroso derivatives of 2-NA and *N*-hydroxy derivatives of 4-ABP, which are metabolites of each aromatic amine, cause oxidative DNA damage with redox reaction.

Mononuclear compounds, such as benzene and *o*-toluidine, seem to express carcinogenicity by causing oxidative DNA damage rather than DNA adduct formation after

* Tel: +81 (59) 340-0568; Fax: +81 (59) 368-1271; E-mail: kawanisi@suzuka-u.ac.jp

metabolic activation. Oxidative DNA damage can play an important role in the carcinogenic processes of PAHs and aromatic amines, in addition to bulky-DNA adduct formation.

In: Genetics Research Summaries
Editors: L. E. Hewitt and J. W. O'Neil

ISBN: 978-1-62257-646-3

Chapter 51

The Usefulness of Bulky DNA Adduct Formation as a Biological Marker of Exposure to Airborne Particulate Matter (PM2.5) in In Vitro Cell Lung Models: A Comparative Study

Sylvain Billet[1,2,3], *Véronique André*[3], *Imane Abbas*[1,2], *Jérémie Le Goff*[3], *Anthony Verdin*[1,2], *François Sichel*[3], *Pirouz Shirali*[*1,2] *and Guillaume Garçon*[1,2]

[1]Univ Lille Nord de France, Lille, France

[2]Toxicologie Industrielle et Environnementale, Université du Littoral – Côte d'Opale, Dunkerque, France

[3]Groupe Régional d'Etudes sur le Cancer, Université de Caen Basse-Normandie et Centre Régional de Lutte Contre le Cancer François Baclesse, Caen cedex, France

Research Summary

Epidemiological evidence suggests the existence of a relationship between high environmental levels of Particulate Matter (PM) and increased incidence of both morbidity and mortality due to respiratory and/or cardiovascular diseases. The consistency among the findings of epidemiological studies argues for a causal association, but it is still difficult to attribute acute health effects to concentration levels in light of the current knowledge. The mechanisms underlying these adverse effects are also not well understood, and major questions still remain concerning the specific size fraction, chemical composition and

* Corresponding Author: Pr Pirouz Shirali, LCE EA2598, Toxicologie Industrielle et Environnementale, MREI 2, ULCO, 189A, Avenue Maurice Schumann, 59140 Dunkerque, France; Phone number: +33-3-28237610; Fax number: +33-3-28237171; E-mail: Pirouz.Shirali@univ-littoral.fr

causative toxicological mechanisms leading to the observed health effects. Hence, to improve the knowledge of air pollution PM-induced toxicity in human lungs, with a particular interest in the crucial role played by coated organic compounds, we focused our attention on the metabolic activation of Polycyclic Aromatic Hydrocarbon (PAH)-coated onto air pollution PM and, thereafter, the formation of PAH-DNA adducts in three in vitro cell lung models: A549 cell line, L132 cell line and human Alveolar Macrophages (AM). PAH, PolyChlorinated Dibenzo-p-Dioxins and -Furans (PCDD/F), Dioxin-Like PolyChlorinated Biphenyls (DLPCB) and marker PolyChlorinated Biphenyls (PCB) coated onto collected PM were determined (i.e., GC/MS and HRGC/HRMS, respectively). Cells were exposed to Dunkerque City's $PM_{2.5}$ at its lethal concentrations at 10% and 50% (i.e., A549: LC_{10} = 6.33µg/cm²; and LC_{50} = 31.63µg/cm²; L132: LC_{10} = 5.02µg PM/cm²; and LC_{50} = 20.10µg PM/cm²; AM: LC_{10} = 5.97µg PM/cm²; and LC_{50} = 29.85µg PM/cm²). Cytochrome P450 (CYP) 1A1 gene expression (i.e., RT-PCR) and enzymatic activity (i.e., EROD activity) and the formation of PAH-DNA adducts (i.e., ^{32}P-postlabelling) were investigated after 24, 48 and/or 72h. Particular (i.e., desorbed PM, dPM) and PAH-positive (i.e., Benzo[*a*]Pyrene, B[*a*]P; 1µM) controls were included in the experimental design. A statistically significant increase in *CYP1A1* gene expression was observed in the three cell lung models under study, 24, 48 and/or 72h after their exposure to PM or B[*a*]P. Similarly, *CYP1A1* gene expression was highly induced 24, 48 and/or 72h after lung cell exposure to dPM, suggesting thereby that the employed outgassing method was not efficient enough to remove total PAH.

Accordingly, EROD activity was significantly increased only in A549 cells and AM, 24, 48 and/or 72h after their exposure to dPM, PM or B[*a*]P. Even if B[*a*]P exposure caused important bulky DNA adduct formation, only very low levels of PAH-DNA adducts, which could not be reliably quantified, were reported 72h after A549 cells and AM exposure to PM and dPM. The relatively low levels of PAH, together with the presence of PCDD/F, DLPCB and PCB coated onto Dunkerque City's $PM_{2.5}$ could contribute to explain this borderline detection of PAH-DNA adducts. Neither significant EROD activity nor PAH-DNA adduct formation could be measured in PM- or in B[*a*]P-exposed L132 cells. In contrast, A549 cells and AM seem to present higher sensitivity to coated organic chemical-induced damage. We also concluded that, in the three lung cellular models we used and in the experimental conditions we chose, bulky DNA adduct formation only partly contribute to the lung toxicity arising from the exposure to Dunkerque City's $PM_{2.5}$.

In: Genetics Research Summaries
Editors: L. E. Hewitt and J. W. O'Neil
ISBN: 978-1-62257-646-3

Chapter 52

SV40 and Cancer

***Maria E. Ramos-Nino*[1] *and Maurizio Bocchetta*[2]**
[1]Department of Pathology,
University of Vermont College of Medicine, Burlington, Vermont, US
[2]Thoracic Oncology Program, Cardinal Bernardin Cancer Center,
Loyola University, Maywood, IL, US

Research Summary

Simian Vacuolating virus 40 (SV40) is a small DNA oncogenic virus first isolated in 1960 from contaminated polio vaccines. Soon after its discovery, it was recognized SV40 is highly oncogenic in experimental animals. The types of tumors induced by SV40 depend on the route of viral administration. For this reason, SV40 is classified as polyomavirus. The SV40 genome contains a limited amount of genetic information. Hence, this virus depends almost entirely on the host's biochemical machinery for replication. Nevertheless, SV40 employs specific strategies to deregulate the host's cell cycle to replicate its own genome. SV40 achieves this task through many elegant molecular biology strategies which allow it to maximize usage of the SV40 circular chromosome. More importantly, all protein products coded for the 5-6 SV40 genes are highly multifunctional during each phase of the viral infection. The degree of SV40 proteins' multifunctionality has been extraordinarily exploited in the viral major oncogene (the Large T antigen, or Tag), which interacts with (and influences the activities of) a large number of cellular proteins. Throughout the last four decades, SV40 has been used extensively as a tool for understanding basic molecular biology concepts; yet, this apparently simple virus always reserves new surprises for researchers. In the past decade, either alone, or in cooperation with environmental carcinogens, SV40 has been increasingly linked to human carcinogenicity. Interest in SV40 has increased because of its presence in certain forms of human cancers, including mesothelioma, brain, bone tumors, and more recently, in non-Hodgkin's lymphoma. However, though evidence of SV40's presence has been found in certain tumors, it has not been determined SV40 actually causes these cancers. Population-based studies are conflicting, and more are necessary to address causality. Many epidemiological studies found no increased risk for developing cancer among individuals exposed to SV40-contaminated vaccines; however, all of these studies had significant limitations, including: small sample size of rare tumors, ambiguity in defining exposed and control individuals, and technical difficulties. These limitations make it difficult to draw any definitive conclusions on the increased risk of human cancer associated with SV40-contaminated polio vaccine. Nevertheless, the presence of SV40 DNA sequences in individuals not directly exposed to SV40-contaminated vaccines suggests there may be other modes of viral transmission in human populations which could confound these results. This review presents some basic aspects of SV40's oncogenic potential; its role in cancer; the epidemiological evidence shaping controversies in the field; present efforts; and a general view on the potential future directions.

In: Genetics Research Summaries
Editors: L. E. Hewitt and J. W. O'Neil
ISBN: 978-1-62257-646-3

Chapter 53

New Insights into the Role of the dUTPase in Epstein-Barr Virus Replication and Pathogenesis

***Marshall Williams*[1,2,*] *and Ronald Glaser*[1,2,3]**
[1]The Department of Molecular Virology, Immunology and Medical Genetics,
[2]The Comprehensive Cancer Center, The Ohio State University Medical Center
[3]The Institute for Behavioral Medicine Research,
The Ohio State University Medical Center, Columbus, OH, US

Abstract

Epstein-Barr virus (EBV), a gamma herpesvirus, is implicated in the pathogenesis of a variety of human malignancies including Burkitt's lymphoma (BL), nasopharyngeal carcinoma (NPC), Hodgkin's disease, and post-transplant B-cell lymphoma. Studies examining the mechanisms by which EBV causes malignancies have focused on those viral encoded proteins that are expressed during latency but there have been few studies to determine the potential role that proteins produced during lytic replication of EBV may have in the pathophysiology of EBV-associated diseases. We have demonstrated that the EBV-encoded deoxyuridine triphosphate nucleotidohydrolase (dUTPase), an enzyme that is produced during lytic replication of EBV, causes immune dysregulation in addition to its enzymatic function. The EBV-encoded dUTPase induces the upregulation of several proinflammatory cytokines including TNF-α, IL-1β, IL-6 and IL-8 as well as IL-10 in human monocytes/macrophages through the activation of NF-κB. Furthermore, the EBV-encoded dUTPase inhibited human T-cell proliferation in vitro. The EBV-encoded dUTPase also inhibited the replication of mitogen-stimulated lymphocytes and the synthesis of interferon-γ

*Correspondence concerning this article should be addressed to: Marshall Williams, PhD, The Ohio State University, Department of Molecular Virology, Immunology and Medical Genetics, 2074 Graves Hall 333 W. 10th Ave, Columbus, OH 43210, Phone: (614) 292-0717; Fax: (614) 292-9805; e-mail: Williams.70@osu.edu.

by cells isolated from lymph nodes and spleens of mice treated with the dUTPase. The data provide a new perspective on how a nonstructural protein that is associated with lytic replication of EBV can cause immune dysregulation. While the production of these proinflammatory cytokines may contribute to the pathology associated with EBV infection, perhaps more importantly, the production of IL-10 may alter the response of T-cells, ($CD4^+$ and $CD8^+$) to EBV infection. This may create a favorable environment for controlling the steady state expression of latent EBV and the ability of EBV genome positive tumor cells to survive.

In: Genetics Research Summaries
Editors: L. E. Hewitt and J. W. O'Neil
ISBN: 978-1-62257-646-3

Chapter 54

EBV Virus and Cancer

Viroj Wiwanitkit
Wiwanitkit House, Bangkhae, Bangkok, Thailand

Research Summary

Epstein Barr virus (EBV) is a DNA virus at can be the cause of several cancers. This DNA virus is proved as DNA tumor virus. Examples of EBV – related cancers include nasopharyngeal carcinoma and lymphoma. The reports on oncogenesis of EBV in significant cancers will be discussed in this chapter. Detection for the EBV is a commonly applied virological laboratory. In this work, the present novel knowledge on EBV laboratory diagnosis, therapy and prevention are reviewed and discussed. In addition, the author also performs a metanalysis on the previous reports on EBV virus infection in tropical countries.

In: Genetics Research Summaries
Editors: L. E. Hewitt and J. W. O'Neil
ISBN: 978-1-62257-646-3

Chapter 55

DNA Tumor Viruses: Oncogenesis of Human Papillomavirus (HPV)

Masachika Senba[1], Naoki Mori[2] and Akihiro Wada[3]
[1]Department of Pathology, Institute of Tropical Medicine, Nagasaki University, Nagasaki, Japan
[2]Division of Molecular Virology and Oncology, Gradual School of Medicine, University of the Ryukyus, Okinaw, Japan
[3]Department of Microbiology, Institute of Tropical Medicine, Nagasaki University, Nagasaki, Japan

Research Summary

Human papillomaviruses (HPV) have been etiologically associated with malignant lesions, most notably cervical, penile, anal and oro-pharynxal. Approximately 40 HPV types have been associated with lesions of the anogenital tract and these can be further classified into so-called low-risk and high-risk types based on their association with clinical lesions. While low-risk HPVs are generally associated with benign lesions such as condyloma acuminatum, infections with high-risk HPVs are associated with a significantly increased risk of developing cancer. Based on epidemiological and moleclular biological evidence, it is widely accepted that high risk HPVs, including HPV types 16 and 18, play an etiological role in the development of cancer. Two oncogenic viral proteins, E6 and E7, involved in cell cycle control, are required for the efficient immortalization of their natural host cells, resulting in primary human squamous epithelial cells developing mucosal and skin cancers. These proteins have been the focus of many studies of cervical cancer during the past 20 years, and their oncogenic mechanisms are now established E7 binds to and degrades the tumor suppressor Rb, while E6 binds to and inactivates the tumor suppressor p53. The causal relationship between chronic inflammation and cancer is widely accepted. Specifically, there is a strong association between tumor viruses and development of cancers. Activation of NF-kB is also involved cancer development and progression. Therefore, this short review focuses on the molecular players during the development from chronic inflammation to cancer.

In: Genetics Research Summaries
Editors: L. E. Hewitt and J. W. O'Neil

ISBN: 978-1-62257-646-3

Chapter 56

Present focus on HPV virus Generated Carcinoma

Viroj Wiwanitkit
Wiwanitkit House, Bangkhae, Bangkok Thailand

Research Summary

Human papilloma virus (HPV) is a DNA virus that can be the cause of many carcinomas. This DNA virus is proved as DNA tumor virus. The cervical carcinoma is the main carcinoma that relates to HPV infection. Screening for the HPV DNA is the new laboratory analysis in gynecology. Also, the vaccine for HPV is presently available. In this work, the present new knowledge on HPV laboratory diagnosis, therapy and prevention are reviewed and discussed. In addition, the author also performs a metanlysis on the previous reports on cervical carcinoma, the most common female carcinoma in Thailand, and HPV virus infection in Thailand.

In: Genetics Research Summaries
Editors: L. E. Hewitt and J. W. O'Neil
ISBN: 978-1-62257-646-3

Chapter 57

Gammaherpesviruses and Oncogenesis*

M. Kúdelová and J. Rajčáni
Institute of Virology, Slovak Academy of Sciences,
Bratislava, Slovak Republic

Research Summary

Gammaherpesviruses comprise a subfamily (*Gammaherpesvirinae*) of the *Herpesviridae* genus; they fall into two genera: Rhadinovirus and Lymphocryptovirus. Human gammaherpesviruses encompass the Epstein-Barr virus (EBV, HHV-4) and the Kaposi´s sarcoma herpesvirus (KSHV, HHV-8). Their genomes as well as the genomes of several mammalian gammaherpesviruses were fully sequenced. The gammaherpesvirus genes may be grouped according to several criteria. As a rule, non-structural and structural genes can be distinguished. Another commonly used classification is related to transcription kinetics, distinguishing immediate-early, early, early-late and late genes. Genes which expression is inevitable for productive virus replication *in vitro* are usually referred to as essential ones. The non-structural early gammaherpesvirus genes encode enzymes for viral DNA synthesis and specify polypeptides involved in immune evasion and in complement regulation. These also modulate or hamper the effects of cytokines and/or disregulate interferon action. During latency, which can be defined as non-productive maintenance of the circularized episomal dsDNA genome, a few (or at least one) latency-associated gene(s) are (is) expressed. Comparative sequence analysis of herpesvirus genomes throughout all subfamilies revealed the group of herpesvirus common genes (showing relatively wide sequence homologies), which encode structural as well as non-structural proteins with related functions. An important group of genes was found gammaherpesvirus specific (γ-specific). Finally, about 10

* A version of this chapter was also published *in Oncogenic Viruses Research Trends*, edited by Lilian T. Johannes, published by Nova Science Publishers, Inc. It was submitted for appropriate modifications in an effort to encourage wider dissemination of research.

– 15 genes may be unic for each gammaherpesvirus. Such genes, present in the given gammaherpesviruses only, may encode proteins, which act in analogical manner though being different in their sequence. The products of gammaherpesvirus specific genes are often virus-coded counterparts of cellular genes; such proteins usually promote cell division and/or cause false intracellular signaling. The latter are believed to represent pirated cellular genes, which have been modified by evolution when becoming virus genome components. Viral oncogenes and virus-coded immune regulation proteins act in accord with infected lymphocytes. Since gammaherpesviruses remain latent in lymphatic tissues and may be associated with lymphoproliferation, comparison of their DNA sequences and functional analysis of corresponding proteins are of special interest.

In: Genetics Research Summaries
Editors: L. E. Hewitt and J. W. O'Neil
ISBN: 978-1-62257-646-3

Chapter 58

Human Herpesvirus-8 and Corticosteroids

Celeste Lujan Pérez and Mónica Isabel Tous
ANLIS Dr. Carlos G. Malbrán, Buenos Aires, Department of Virology,
National Institute for Infectious Disease, Argentina

Research Summary

Kaposi´s sarcoma (KS) associated herpesvirus (KSHV), also termed human herpesvirus 8 (HHV-8) was discovered in 1994 when Chang and colleagues reported the finding of herpesvirus-like sequences in skin lesions from AIDS-KS patients [1]. The virus is present in all KS clinical forms, and it is also associated with another two neoplasic diseases: multicentric Castleman's disease (MCD) and primary effusion lymphoma (PEL). PEL is a rare case of lymphoma first identified as a subset of body-cavity-based lymphomas BCBL [2]. These lymphomas present initially as a lymphomatous effusion located in body cavities without a contiguous tumor mass. PEL cells can be *in vitro* cultured and cell lines PEL-derived are a well-characterized model for a cancer caused by the expression of a few HHV-8 latent proteins [3-5]. Even though most PEL cells are co-infected with HHV-8 and Epstein Baar virus (EBV), some lines, such BC-3, KS-1, BCP-1, CROAP/3, and BCBL-1 contain only HHV-8 episomals [5-9]. Episomes can turn from latent to viral progenie after stimulation with phorbol ester 12- O – tetradenoyl phorbol-13 acetate (TPA), enlarging 16 times the amount of virus [10]. Serum antibodies that are bound to the virus sprouting in the cell are revealed by indirect immunoflourescence assay (IFA). The use of corticosteroids in transplanted patients increases the occurrence and severity of KS [11]. *In vitro,* hydrocortisone induces a viral lytic reactivation in BCBL-1 cell line [12]. In our laboratory, we have examined the ability of dexamethasone (DEX) to induce HHV-8 lytic phase on BCBL-1. We compared the action of DEX with TPA on cell viability, amount of antigen expression, and the performance of IFA slides to detect antibodies against HHV-8 in KS patients, HIV infected individuals and blood donors. Results were analysed by Fisher exact test. Viable cells were counted daily by trypan blue exclusion for six days. Differences between TPA-induced and control cells were found, but no differences were seen in DEX vs. control cells. Daily viral antigen expression within the cells was examined by IFA using HHV-8 positive pool antisera [13]. Slides containing TPA, or DEX-induced cells were made up and cells expressing the HHV-8 lytic pattern were counted by IFA [14]. No differences (p=0.67) were observed among the amount of fluorescent cells in slides with TPA-or DEX-induced cells. Also, no differences were observed between TPA and DEX on fluorescent pattern. In order to investigate if DEX induced cells may be used for HHV-8 serological diagnosis, 86 HHV-8 positive sera from previously described populations were IFA tested and titered in TPA and DEX slides; 54 HHV-8 negative sera were also included [14]. Results fully matched both slides. Our data demonstrate that DEX could be used for studies of HHV-8 lytic cycle activation events and as an alternative to TPA induction to make HHV-8 IFA slides for serologic diagnosis.

In: Genetics Research Summaries
Editors: L. E. Hewitt and J. W. O'Neil

ISBN: 978-1-62257-646-3

Chapter 59

Polyomavirus and Cancer

Romina Bonaventura* and María Cecilia Freire
Neurovirosis Division, Department of Virology,
National Institute for Infectious Disease,
ANLIS Dr. Carlos G. Malbrán, Buenos Aires, Argentina

Research Summary

The polyomavirus family includes two members, BK virus (BKV) and JC virus (JCV), that naturally infect humans. Besides these, the primate polyomavirus, simian vacuolating virus (SV40), was isolated from humans as a consequence of contaminated polio vaccines in the 1950's in theUnited States. Human polyomaviruses are ubiquitous in the human population worldwide and cause disease only in immunocompromised individuals: progressive multifocal leukoencephalopathy (PML) mainly in AIDS patients and polyomavirus-associated nephropathy (PVAN) in renal transplant patients, being the etiologic agent JCV and BKV, respectively. Many studies indicate that these three polyomaviruses are associated with cancers in humans as their genome sequences and protein expression are detected in several types of human neoplasm. Polyomaviruses rely on cellular enzymes and cofactors for DNA replication and these proteins are expressed in S phase; then JCV, BKV and SV40 encode regulatory proteins (large T antigen, middle T antigen and small t antigen) that drive resting cells into S phase through interactions with cell cycle regulators such as pRb and p53. This aberrant stimulation of the cell cycle is a driving force of oncogenic transformation. Here we describe what is known until now about the molecular mechanisms of these interactions.

* E-mail: rbonaventura@anlis.gov.ar

In: Genetics Research Summaries
Editors: L. E. Hewitt and J. W. O'Neil

ISBN: 978-1-62257-646-3

Chapter 60

Mechanisms by which Viral Proteins of Human DNA Tumor Viruses Hijack the Cell Cycle

Mona Johannessen and Ugo Moens *
University of Tromsø, Faculty of Medicine, Institute of Medical Biology,
Department of Microbiology and Virology, Tromsø, Norway

Research Summary

Viruses are obligate intracellular parasites with a restricted genome size encoding a limited number of genes. To compensate for their small number of proteins, viruses often usurp cellular proteins to complete their life-cycle. Their propagation time in a permissive host cell is often very short and to speed up their replication, these viruses can subvert the host cell cycle. Despite their vast genetic diversity, DNA tumor viruses have evolved common mechanisms to disorder the cell cycle. Viral-encoded proteins can activate cellular proteins that allow the cell to enter the S phase or to arrest in the G_2 phase. Moreover, viral proteins may initiate cell cycle progression by destroying the suppressive activity of tumor suppressors such as p53, the pRb family members, and cyclin-dependent protein kinase inhibitors. In addition, some of these viruses encode their own cyclins that allow them to steer cellular cyclin-dependent kinases. This review will focus on the different mechanisms by which human DNA tumor viruses hijack the cell cycle, and will discussed the application of cell cycle inhibitors as potential therapeutic against viral infections.

* Corresponding author: Ugo Moens, University of Tromsø, Faculty of Medicine, Institute of Medical Biology, Department of Microbiology and Virology, N-9037 Tromsø, Norway. Phone: +47-77644622, fax: +47-77645350, e-mail: ugom@fagmed.uit.no.

In: Genetics Research Summaries
Editors: L. E. Hewitt and J. W. O'Neil
ISBN: 978-1-62257-646-3

Chapter 61

The Immune Response to Tumor-Associated Antigens in Hematological Malignancies – Its Relevance for Vaccine Development

Kamel Ait-Tahar** *and Karen Pulford
University of Oxford, Nuffield Department of Clinical Laboratory Sciences,
John Radcliffe Hospital, Oxford, UK

Research Summary

Study on the immune response to tumor-associated antigens (TAAs) has advanced significantly over the last twenty years. The aim of such investigations has been the identification of molecules that may be targeted to cause not only regression of the tumor but also its eradication. Such TAAs may represent molecules that are expressed solely by the tumour cells while other TAAs are more widely distributed in normal cells. Spontaneous cytotoxic T cell activity and the presence of circulating antibodies, the latter suggestive of a cognate CD4 T cell response, have been reported in some malignancies. The immune response was, however, insufficient to control tumor growth possibly as a result of T cell tolerance. Strategies are now being developed to produce vaccines that are capable of optimising the immune recognition of the TAAs to overcome the problems of tolerization but without eliciting major autoimmune reactions.

In this review, we will enumerate some of the major TAAs studied in the context of hematological malignancies. We will outline the techniques used to identify these molecules and assess their usefulness in immunotherapy. Vaccination-based therapies where both T cell- and antibody-mediated anti-tumor responses are likely to be relevant will also be explored. As

* Correspondence concerning this article should be addressed to: Dr Karen Pulford, Nuffield Department of Clinical Laboratory Sciences, Level 4, Academic Block, John Radcliffe Hospital, Oxford, OX3 9DU, U.K. Tel: 44 1854 220246; Fax: 44 1865 220078; Email: karen.pulford@ndcls.ox.ac.uk.

an example, we will highlight our studies on the immune responses to the lymphoma-associated anaplastic lymphoma kinase (ALK) protein expressed in the majority of cases of anaplastic large cell lymphoma. We will show the potential of the ALK protein as an immunotherapeutic target since evidence for both humoral (antibody) as well as cellular (CTL and CD4 T-helper) responses has been obtained.

In: Genetics Research Summaries
Editors: L. E. Hewitt and J. W. O'Neil

ISBN: 978-1-62257-646-3

Chapter 62

The Search for Human Tumor Viruses

Herbert Bosshart*
Department of SurgeryZurich, and Department of Psychiatry, Zurich,
University Hospital, Switzerland

Research Summary

Viruses are thought to contribute to the development of at least 15% of all human cancers. The ubiquitous human papilloma viruses are responsible for more than 80% of cervical cancer-related deaths. Hepatitis B and hepatitis C viruses are found worldwide and are associated with the majority of liver cancers. Epstein-Barr virus is strongly linked to Burkitt lymphoma, the most frequently occurring childhood cancer in Africa. The same virus is also associated with the development of nasopharyngeal carcinoma, which is prevalent in Southeast Asia. The human T-lymphotropic virus-1 is endemic in South Japan and Central Africa and infected individuals carry the risk of developing adult T cell lymphocytic leukemia. The successful development of vaccines against some of these viruses illustrates the importance of continuing to search for novel viruses in samples from tumors with suspected viral etiologies. Global amplification of polyadenylated transcripts combined with serial analysis of gene expression can be used to analyze the transcriptome of single cells in an unbiased fashion. Using this approach, it should be possible to detect rare viral transcripts in latently infected tumor cells. The long-held hypothesis that a number of human malignancies, including certain types of childhood leukemia and lymphoma, may have underlying viral etiologies, could be tested in this way.

* Correspondence: Herbert Bosshart, MD, Zurich University Hospital, Department of Psychiatry, Lenggstrasse 31, P. O. Box 68, CH-8029 Zurich, Switzerland. Phone: ++ 41 1 384 2111 ; Fax: ++ 41 1 383 4456 ; E-mail address: herbert.bosshart@usz.ch

In: Genetics Research Summaries
Editors: L. E. Hewitt and J. W. O'Neil

ISBN: 978-1-62257-646-3

Chapter 63

Macrophage Involvement in the Inflammation-Induced Onset of Epstein-Barr Virus-Related Tumor

Misuzu Shimakage[1] and Haruhiko Sakamoto[2]
[1]Department of Pediatrics, National Hospital Organization,
Wakayama National Hospital, Wada, Mihama-cho, Hidaka-gun, Japan
[2]Department of Pathology, Faculty of Medicine, Kagawa University,
Ikenobe, Miki-cho, Kita-gun, Japan

Research Summary

Epstein-Barr virus (EBV) is known as a causative agent of Burkitt's lymphoma, nasopharyngeal carcinoma, and about 10% of the cases of stomach carcinoma. We have reported EBV gene expression including EBV lytic infection protein in many other human cancers. We have also identified the expression and replication of EBV genes in normal human macrophages, and abnormal histiocytes in Langerhans' cell histiocytosis (LCH). Furthermore, we revealed EBV expression in macrophages in the lesions of LCH, primary lung lymphoma, uterine cervical carcinoma, renal cell carcinoma, tongue cancer, cutaneous T-cell lymphoma, and thyroid carcinoma. Therefore, tumor-associated macrophages in EBV-related tumors are likely to carry EBV, and they appear to induce EBV lytic infection. Lytic infection of EBV in multiple tissues may provoke a strong inflammatory response, because cell lysis induced by virus replication results in intense immune responses against viral proteins. We reported that childhood LCH expressed high level of EBV lytic infection protein, and that, in one case, the administration of acyclovir resulted in complete remission. Anti-herpesvirus drugs seem to be effective against the disease. We reported that uterine cervical carcinoma expresses EBV genes which participate in transformation, and that carcinoma cells of almost all cases of cervical carcinoma also carried the HPV 16 gene. It is interesting that EBV and HPV may cooperate in the development of cervical carcinoma through the long-term inflammation caused by chronic infection with both viruses. The roles of macrophages not only in inflammation but also in the interaction between viruses or viruses and microbes should be clarified. Through such studies of inflammation involving macrophages, the dynamic mechanism leading to EBV oncogenesis will be elucidated.

In: Genetics Research Summaries
Editors: L. E. Hewitt and J. W. O'Neil
ISBN: 978-1-62257-646-3

Chapter 64

Acquired Antibiotic Resistance in Starters and Probiotics Strains of Lactic Acid Bacteria and Bifidobacteria Species

Baltasar Mayo**, Ana B. Flórez, Mohammed S. Ammor and Susana Delgado

Department of Microbiology and Biochemistry, Instituto de Productos Lácteos de Asturias (CSIC), Villaviciosa, Asturias, Spain

Research Summary

Lactic acid bacteria (LAB) are non-pathogenic microorganisms found naturally in raw plant and animal materials or added deliberately as functional starter cultures to promote desirable biochemical changes in food and feed fermentations. LAB and bifidobacteria are also natural inhabitants of the gastrointestinal tract (GIT) of human and animals where they exert an array of protective and beneficial effects; a property that has led to the use of members of these two microbial types as probiotics. Antibiotics have greatly contributed over the last 65 years to human life expectancy and quality of life in the fight against infectious diseases. However, their efficacy is being severely threatened by the appearance and worldwide spread of antibiotic resistance. At present, there is great concern that commensal and beneficial bacterial populations, such as the foodborne and GIT-associated bacteria, become a reservoir for resistances, from which they could ultimately be transferred to human pathogens and opportunistic bacteria. Transferable resistances carried by starters and probiotics could be spread either during food manufacture or during passage through the GIT.

* Corresponding Author: Baltasar Mayo, Department of Microbiology and Biochemistry, Instituto de Productos Lácteos de Asturias (CSIC), Carretera de Infiesto s/n, 33300 Villaviciosa, Asturias, Spain Phone number: +34 985 89 12 31 Fax number: +34 985 89 22 33 E-mail address: baltasar.mayo@ipla.csic.es

Consequently, discriminating between resistant and susceptible strains, and distinguishing between intrinsic, nontransferable and acquired, transferable resistances is vital in order of not to disseminate antibiotic resistances via the food chain. Genes conferring resistance to several antibiotics have already been identified and characterized in strains of LAB and bifidobacteria. In particular, *tet*(W) and *erm*(B), encoding tetracycline and erythromycin resistance, respectively, seem to be widely spread among species of these bacterial groups. Some of these genes have proved to be transferable in laboratory conditions, but also in food and in animal GIT systems.

In: Genetics Research Summaries
Editors: L. E. Hewitt and J. W. O'Neil
ISBN: 978-1-62257-646-3

Chapter 65

Antibody Variable-Region Sequencing for Hybridoma Authentication

Miha Kosmač[1,2], Simon Koren[1,2], Nives Škrlj[2], Marko Dolinar[2] and Vladka Čurin Šerbec[1,2]

[1]Department for Production of Diagnostic Reagents and Research, Blood Transfusion Centre of Slovenia, Ljubljana, Slovenia
[2]Faculty of Chemistry and Chemical Technology, University of Ljubljana, Ljubljana, Slovenia

Research Summary

Miss-identification and cross-contamination of cell lines used in research and biotechnology represent widespread problems. Appropriate cell authentication methods should thus be used regularly to avoid invalidating scientific conclusions and to assure the quality of biotechnological products. DNA fingerprinting is often used to establish and verify the identities of various cell lines, but this approach is unsuitable for hybridoma cell lines that are derived from syngeneic animals. Here, we describe the sequencing of antibody variable regions, the only highly diverse region of the hybridoma genome, as a robust and accessible method for hybridoma cell line authentication. The protocol involves RNA isolation, reverse transcription of immunoglobulin variable regions, and amplification of the resulting cDNA using highly degenerate primers, chosen to amplify the majority of the possible variable regions. Depending on the quality of PCR amplification, the PCR products can then be sequenced either directly or after cloning into a plasmid vector. Additionally, we provide an alternative protocol based on rapid amplification of cDNA ends, which can be used to obtain variable region sequences where successful amplification with the degenerate primers is not achieved.

In: Genetics Research Summaries
Editors: L. E. Hewitt and J. W. O'Neil

ISBN: 978-1-62257-646-3

Chapter 66

DNA Sequencing: Methods, Strategies and Protocols

***M. Naum*[1], *R. Bell*[1], *M. Mammel*[2], *J. Zheng*[1] *and E. W. Brown*[1]**
[1]Food and Drug Administration,
Division of Microbiology, College Park, MD, US
[2]Food and Drug Administration,
Division of Molecular Biology, Laurel, MD, US

Research Summary

DNA sequencing is a common and requisite practice for molecular biologists today working in all areas of biology, including microbiology. A number of nucleotide and amino acid-based sequence typing methods are now widely recognized and have transformed the way in which complex evolutionary and epidemiological relationships are elucidated among bacterial strains. Recent advances in DNA sequencing underpin a vast array of bacterial subtyping methods and have become an integral component to traceback investigations of foodborne outbreaks caused by specific bacterial pathogenic strains. Direct DNA sequence-based approaches include multi-locus sequence analysis (MLSA), which simultaneously analyzes the nucleotide substitutions found within concatenated sequence alignments from multiple housekeeping genes. This collective and simultaneous analysis of nascent single nucleotide polymorphisms (SNPs) is effective in differentiating highly homogeneous strain clusters including the closely related strains and serovars of *Salmonella* enterica. In addition to the extraordinary power this approach offers in molecular evolution/epidemiology, the study of concatenated gene sequences has been instrumental in defining a new paradigm for horizontal gene transfer in the genetic structure of many bacterial species. Facile variations of MLSA are now being developed that rely on the culling of SNP-based targets from relatively uncharted and under-exploited regions of the bacterial genome. As an example, SNPs that reside in the selectively neutral intergenic non-transcribed spacer regions of the genome hold great promise in resolving bacterial strains down to a level of epidemiological relevance. In parallel with these advances in nucleotide sequence analysis, the development of rapid and highly accurate SNP discovery and detection tools is and includes methods such as Pyrosequencing, Luminex probe recognition technology, and single-basepair primer extension assays to name but a few. Together, these successes clearly position DNA sequence-based strategies at the helm for future discoveries in the genetics, evolution, and epidemiology of bacterial strains and species.

In: Genetics Research Summaries
Editors: L. E. Hewitt and J. W. O'Neil

ISBN: 978-1-62257-646-3

Chapter 67

DNA Sequencing: Methods, Strategies and Protocols in Molecular Biology Research Horizons

***Cimaglia Fabio*[1], *Assab Emanuela*[1], *D'Urso Oscar Fernando*[1] *and Poltronieri Palmiro*[*2]**

[1]Department of Environment and Biology Sciences, Lecce, Italy
[2]CNR, Institute of Sciences of Food Productions, Lecce, Italy

Research Summary

In recent years the exploitation of sequenced genomes has made possible the deepening of knowledge on how many genes are contained in the genomes of higher organisms. The identification of thousands of functional RNAs showed that sequenced genomes contain much more genes than previously sought. We cloned a library of RNAs sized 60-500 bases mouse and identified thirty small RNAs isolated from the developing embryo brain, the major number of them belonging to H/ACA and C/D box snoRNAs. Many of these sRNAs and snoRNAs are coded in introns of protein coding and nonprotein coding transcripts. The small RNAs can form secondary structures with free energy ranging from -3.4 to -70 kcal/mol. Three-dimensional architectural motifs are increasingly recognized as determinants of RNA functionality. Such motifs can encode spatial information required for interaction with biomolecules. Localisation on the mouse genome using the UCLA Santa Cruz server showed a high conservation in these short sequences with overlapping regions of other genomes. Most of these new short RNAs have been identified today with an ENSEMBL identification number, but in our sequences there are 5' or 3' ends differences, probably relative to processing events and enzyme modifications.

[*] Correspondence author: Palmiro.poltronieri@ispa.cnr.it, Tel. +390832422609

A different DNA sequencing approach was used to identify genes from organisms with unsequenced genomes of wild species related to cultivated crops. In Solanaceae, gene duplication events produced a highly variable number of sequences coding for protein inhibitors targeting proteases, hydrolases and polygalacturonases. Exploiting degenerate primers and PCR amplification, we identified new Kunitz-type proteinase inhibitors of group A, group B and group C from wild Solanum species (*S. palustre* and *S. stoloniferum*), as well as additional isoforms from potato varieties. The new data allowed to design a phylogenetic tree grouping all known Kunitz-type inhibitors in Solanum species. The tree sub-roots, grouping highly related sequences conserved either in *S. palustre* and in *S. tuberosum*, could be useful as marks of gene duplication events at the basis of Solanum sub-families evolution and divergence. This book chapter aims to provide clues for DNA sequencing projects directed to unsequenced organisms in which many transcripts wait to be discovered, either coding small RNAs or genes homologue to known genes coding for protein products.

In: Genetics Research Summaries
Editors: L. E. Hewitt and J. W. O'Neil

ISBN: 978-1-62257-646-3

Chapter 68

RetroArray—A Comprehensive Diagnostic DNA Chip for Rapid Detection and Identification of Retroviruses, Retroviral Contaminants, and Mistaken Identity of Cell Lines

Wolfgang Seifarth[1], *Oliver Frank*[1], *Julia Schreml*[1] *and Christine Leib-Mösch*[1,2]
[1]III. Medizinische Klinik, UMM – Universitätsmedizin Mannheim, University of Heidelberg, Germany
[2]Helmholtz Zentrum München, German Research Center for Environmental Health, Institute of Virology, Neuherberg, Germany

Research Summary

Retroviruses not only represent some of the most dangerous pathogens, but also constitute, as remnants of former infections that happened millions of years ago, a large fraction of the human genome. We have established a fast and reliable DNA chip-based assay (RetroArray) for detection and identification of a wide variety of human and other vertebrate exogenous and endogenous retroviruses in biological/clinical samples. The assay combines a pan-retrovirus multiplex polymerase chain reaction (PCR) using fluorochrome-modified primers and DNA chip hybridization. Using RetroArray, distinct transcription profiles of human endogenous retroviruses (HERVs) have been established for a variety of human tissues. Using paired samples (normal vs. disease) this method can be applied to examine HERV activity in human tumors and can help to identify retrovirus-derived tumor antigens. In addition, RetroArray has been designed to detect human exogenous retroviruses such as human immunodeficiency virus (HIV) and human T-cell leukemia virus (HTLV), and putative cross-species transmissions of animal retroviruses in human samples. Species specificity and sensitivity of the assay was demonstrated by detecting traces of pig endogenous retrovirus (PERV) DNA down to ~25 copies in human cDNA samples. Furthermore, retroviral transcripts may be identified in particle preparations from cell culture supernatants. This makes the assay a valuable technique for monitoring packaging cell lines and vector preparations commonly used in human gene therapy applications to exclude cotransfer of replication competent retroviruses (RCRs) or endogenous retroviruses (ERVs) into target cells. Therefore, RetroArray could improve significantly the safety of human gene therapy, tissue engineering, xenotransplantation and production of therapeutic polypeptides in cell culture. Mistaken identity of human cell lines and frequently observed laboratory contaminations with cells of other species, as well as infection with polytropic animal retroviruses such as squirrel monkey retrovirus (SMRV) or murine leukemia virus (MLV) can influence experimental results and may lead to invalid conclusions. The RetroArray technique is an excellent tool for testing purity and homogeneity of cell lines. Characteristic HERV transcription profiles can be used to assess the cell type and to monitor cell lines for contaminating cells.

In: Genetics Research Summaries
Editors: L. E. Hewitt and J. W. O'Neil
ISBN: 978-1-62257-646-3

Chapter 69

Biogenomic Pathways in Breast and Ovarian Malignancies: Clinical and Surgical Workflow Focused on Tailoring Procedures

A. Tinelli*[*1], A. Malvasi[1], V. Lorusso[2], E. Chiuri Vincenzo[2], S. Cinieri[3], D. Vergara[4], R. Martignago[5], U. Vergari[6], M. Pisano[7] and G. Leo[7]

[1]Department of Gynecology and Obstetric, Vito Fazzi Hospital, Lecce, Italy
[2]Department of Oncology, Oncological Hospital, Lecce, Italy
[3]Department of Oncology, Perrino Hospital, Brindisi, Italy
[4]National Nanotechnology Laboratory (NNL), CNR-INFM, University of Salento, Leece, Italy
[5]Department of Biological and Environmental Sciences and Technologies, Laboratory of Human Anatomy, University of Salento, Leece, Italy
[6]International Center of Bioethics and Human Rights, University of Salento, Lecce, Italy
[7]Molecular Biology and Experimental Oncology Lab, Oncological Hospital, Lecce, Italy

Research Summary

Breast and ovarian cancers are the most common worldwide gynaecological cancers and connected to germline mutations in the tumour suppressor genes BRCA1 and BRCA2, both involved in different cellular mechanism, such as DNA damage recognition, double-strand break repair, checkpoint control, transcription regulation and chromatin remodelling. The inactivation of BRCA-related pathways is probably associated with significantly higher

* Corresponding author: Dr. Tinelli Andrea, MD,Department of Gynecology and Obstetric, Vito Fazzi Hospital, Piazza F. Muratore, 73100, Lecce, Italy, Cellular Number 0039/339/2074078, Fax Number 0039/0832/661511, E-mail: andreatinelli@gmail.com

percentages of these gynaecological cancers. In this elaborate are discussed the biogenomic perspectives in gynecological surgery and the current opportunity to treat breast and ovarian malignancies by tailoring minimally invasive procedures, basing on molecular genetic analysis and by using of prophylactic surgery.

The authors reviewed the clinical and genetic evaluation for germline mutations to monitor the women predisposed to cancer developments and to choose the risk reduction strategies, based on a more precise estimate of individual risk; in the high risk women, the authors focused on the tailoring minimally invasive gynaecological procedures, a preemptive prophylactic surgery based on molecular genetic analysis, targeting to an organ or a part of its prior to malignant transformation or while the cancer is in situ. Clinical and surgical workflow in high risk patients for breast and ovarian cancers include highly sensitive screening, starting at a young age, and the use of tailoring minimally invasive operations on the ovaries or on the breast; the tailoring surgery, based on molecular genetic analysis and targeting a low tumor burden, have the potential to achieve greater cure rates than extensive resections required for clinically evident disease and, moreover, it is as an acceptable surgical approach for smaller incisions, less postoperative pain, and shorter hospital stays.

In: Genetics Research Summaries
Editors: L. E. Hewitt and J. W. O'Neil

ISBN: 978-1-62257-646-3

Chapter 70

Microarray Analysis of Altered Gene Expression in the Kidney of Mice Exposed to Radiation

Mitsuru Nenoi
Radiation Effect Mechanisms Research Group,
National Institute of Radiological Sciences, Inage-ku, Chiba, Japan

Research Summary

The *kidney* is a *radiosensitive organ*, and the radiation-induced nephropathy is a serious problem for cancer patients after clinical radiotherapy. To assess the molecular mechanisms for radiation nephropathy, microarray experiments have been performed in multiple laboratories using RNA from the kidney of mice exposed to radiation. Modulation of a number of genes has been revealed after acute irradiation, including those associated with extracellular matrix synthesis/degradation and regulation of intracellular redox status. However, the risk of radiation nephropathy after chronic exposure is unclear. We previously performed continuous irradiation of C57BL/6J mice with γ-rays for 485 days at dose-rates of 0.032 μGy/min, 0.65 μGy/min, and 13 μGy/min. Gene expression profiles in the kidney from irradiated and unirradiated mice were analyzed, and differentially expressed genes were identified. An evident modulation in expression of the cell cycle-associated genes was not observed in contrast to the case of acute irradiation. However, expression of the genes involved in mitochondrial oxidative phosphorylation was elevated after the long-term irradiation at the dose-rates of 0.65 μGy/min and 13 μGy/min, suggesting that the redox condition in the kidney is likely to be modulated after chronic exposure. Furthermore, it was also found that the genetic response is highly tissue-specific by comparing the gene expression profiles in the kidney and the testis. In this article, analysis of gene expression profile after a long-term irradiation with low-dose-rate γ-rays is reviewed, and the risk of radiation nephropathy after chronic exposure is discussed.

In: Genetics Research Summaries
Editors: L. E. Hewitt and J. W. O'Neil

ISBN: 978-1-62257-646-3

Chapter 71

Methods of Data Analysis Applied to Sequentially Accumulated Gene Signatures

***Shun Ikeda*[1], *Rui Ose*[1, 2] *and Hisashi Koga*[1*]**

[1]Laboratory of Medical Genomics,
Department of Human Genome Research, Kazusa
DNA Research Institute, Kisarazu, Chiba, Japan
[2]Graduate School of Pharmaceutical Sciences & Faculty
of Pharmaceutical Sciences, Chiba University, Chiba, Japan

Research Summary

The DNA microarray is one of the standard assays in modern science. One can obtain tens of thousands of DNA microarray data sets freely through the Internet without performing exact hybridization experiments; however, several different methods of data analysis were used to gather this data, as no standard method has been established. We should therefore focus our effort on data analysis rather than data acquisition. It is promising that important evidence may be extracted from certain sequentially accumulated microarray data. Based on this possibility, we applied three different cluster analyzing methods (UPGMA, SOM, and GSEA) against the sequentially accumulated data from cytokine-treated culture cells and compared the results. In this chapter we briefly review the advantages of each method and look at the usefulness of further projections of the results in pathway databases such as KEGG.

[*] Corresponding Author: Tel: +81 438 52 3505; Fax: +81 438 52 3914; E-mail: hkoga@kazusa.or.jp

In: Genetics Research Summaries
Editors: L. E. Hewitt and J. W. O'Neil

ISBN: 978-1-62257-646-3

Chapter 72

Application of Genomic Fingerprinting Methods for Molecular Characterization of Bioleaching Microorganisms

Bidyut R. Mohapatra**, W. Douglas Gould, Orlando Dinardo and David W. Koren
CANMET Mining and Mineral Sciences Laboratory, Natural Resources Canada, Ottawa, ON, Canada

Research Summary

The acidophilic microorganisms (defined as those which can grow at pH ≤ 3) play a major role in the bioleaching processing of low grade ores for extraction of base and precious metals, including Cu, Zn, Ni, Co, Ag and Au. The bioleaching microorganisms act as the biocatalyst in the oxidation and dissolution of metal sulfide ores by the production of sulfuric acid and ferric ions. The application of bioleaching microorganisms in ore processing has several distinct advantages over conventional mineral processing methods. Microbial extraction processes are believed to be ecofriendly methods owing to the requirement of lesser amounts of energy compared to roasting or smelting, and no production of sulfur dioxide or other gases detrimental to the environment. Bioleaching microorganisms have also attracted significant industrial interest for treatment of toxic heavy metals- and sulfide-containing industrial waste effluents. On the other hand, these microorganisms have a negative impact on public and environmental health with the generation of highly toxic acid mine drainage (AMD). AMD is generated by the subsurface and the surface (open pit) mining of metal ores and coal with the resulting exposure of sulfide minerals to air and water. AMD degrades the quality of surface and groundwater intended for drinking, recreation, crop irrigation and aquaculture with elevated levels of sulfate, and toxic heavy metals and

* Corresponding address: Bidyut Ranjan Mohapatra, #203-Fantable Ninomiya-I, Ninomiya 1-10-19, Tsukuba, Ibaraki 305-0051, JAPAN, Fax: +81-29-861-6733, E. mail: m_bidyut@yahoo.com

metalloids. In order to improve the mineral oxidizing activity of bioleaching microorganisms and to inhibit the production of AMD, it is essential to develop robust genotypic tools for speciation as well as for determination of the biological activities of bioleaching microorganisms. In this chapter, we have provided an overview of the molecular techniques including different genomic fingerprinting methods that have been used for the characterization of bioleaching microorganisms.

In: Genetics Research Summaries
Editors: L. E. Hewitt and J. W. O'Neil
ISBN: 978-1-62257-646-3

Chapter 73

The Use of REP-PCR on Members of the Class *Actinobacteria* Stackebrandt *et al.* 1997

L. A. Maldonado[1] *and E. T. Quintana*[2]
[1]Instituto de Ciencias del Mar y Limnologia (ICMyL) Universidad Nacional Autónoma de México (UNAM) Circuito Exterior s/n Ciudad Universitaria, México DF, México
[2]Escuela Nacional de Ciencias Biológicas Instituto Politécnico Nacional (IPN) Prolongación de Carpio y Plan de Ayala México DF, México

Research Summary

Advances in molecular biology have provided highly specific methods for classifying, identifying and typing isolates based on nucleic acid sequence data. Taxonomic systems are becoming increasingly important for characterizing microorganisms since they can be applied in epidemiological investigations of disease outbreaks to identify a potential common source and/or to establish grouping strategies for isolates recovered from several environments. Typing systems based on phenotypic tests have limitations as key phenotypic traits of microorganisms may be inconsistently expressed or may not provide enough discriminative power to separate closely related microorganisms. Therefore, shortcomings of phenotypically based methods have pushed towards the development and employment of methods based on microbial genotypes or DNA sequences minimising problems with respect to reproducibility and, in some cases, enabling the establishment of large databases (a biological-genotypic database) of characterized organisms. Several molecular typing systems have been used to study the relationships within high G+C bacteria (aka *Actinobacteria*) and the current chapter focuses on the genera *Micromonospora* and *Streptomyces*. Repetitive Extragenic Palindromic Polymerase Chain Reactions (Rep-PCR) fingerprints bacterial genomes based on strain-specific patterns derived from PCR amplification of repetitive DNA elements present within bacterial genomes. The palindromic nature of repetitive elements (ie. BOX, ERIC and REP primers) across the microbial genome and their ability to form stem-loop structures leads to the generation of unique fingerprint patterns. Members of the genus *Micromonospora* cannot be separated on basis of their phenotypic properties, therefore, the use of molecular fingerprinting methods when large numbers of isolates belonging to this genera are urgently needed. The chapter exemplifies its use on nearly 200 micromonosporae recovered from aquatic ecosystems. On the other hand, studies on members of the genus *Streptomyces* -which currently holds nearly 600 species- makes comparative studies difficult, hence also the need for a reliable fingerprinting method. The chapter then deals with studies on streptomycetes isolated from clinical material (11 strains) and their fingerprint relationships. The use of REP-PCR to generate fingerprint patterns and the construction of a biological-genotypic database are provided for these *Actinobacteria* genera.

The accurate circumscription of subtypes within a species is becoming increasingly important in all branches of microbiology. Microbial fingerprinting is extensively used in diagnostic bacteriology, in ecological and evolutionary genetical studies and in search and discovery programmes designed to detect new microbial products. The various molecular fingerprinting methods have advantages and disadvantages when applied to specific situations and objectives. Fingerprinting methods used to discriminate between strains within target species are easy to perform and data interpretation is relatively straighforward. Several molecular fingerprinting systems have been shown to be effective in the delineation of *Actinobacteria* that is Gram positive bacteria of a high Guanine + Citosine content (> 55%), at and below the species level. The introduction and application of molecular taxonomic procedures such as 16S rRNA gene sequencing, DNA fingerprinting DNA: DNA hybridisation, multilocus sequence typing and sequence analyses of complete genomes are providing new insights into prokaryotic systematic, including the classification and identification of the *Actinobacteria.* The development of molecular taxonomic methods based on DNA analyses has made it possible to undertake extensive, rapid and precise characterisation of representatives of bacterial taxa isolated from diverse habitats. Such techniques tend to give results that are more robust than those from chemosystematic studies which are often sensitive to small changes in cultivation conditions. Besides, such techniques provide an insight of the whole genomic information from each strain. Many molecular taxonomic procedures are based on the use of the polymerase chain reaction, which is used to amplify target genes from either culturable isolates or microbial community DNA. The application of different molecular fingerprinting techniques to determine the degree of sequence conservation between bacterial genomes is based on the detection of naturally occurring DNA polymorphisms which are the result of either point mutations or rearrangements in genomic DNA (i.e. insertions or deletions) or fragments of DNA. DNA polymorphisms can be detected by scoring band presence against band absence in banding patterns generated either by restriction enzyme digestion or DNA amplification procedures The underlying premise is that variations in banding patterns are a direct reflection of genetic relationships between the tested strains hence banding patterns can be considered as genomic fingerprints for characterisation and identification purposes. A robust fingerprinting technique can then become a dereplication molecular procedure to indicate whether a specific clinical and/or environmental isolate has been studied (cultured) previously.

In: Genetics Research Summaries
Editors: L. E. Hewitt and J. W. O'Neil

ISBN: 978-1-62257-646-3

Chapter 74

SNPs, Mini-STRs and Molecular Photofitting: Novel Directions for Forensic DNA Typing

Gerhard Mertens
Forensic DNA Laboratory, University of Antwerp,
Antwerp, Belgium

Research Summary

Since Alec Jeffreys in 1985 coined the term "individual-specific genetic fingerprints", DNA typing has become indispensable in forensic analysis, having as central dogma the evidential power of the "match" between stain and suspect. Technological advances in this field include the adoption of the polymerase chain reaction (PCR) to generate DNA profiles from minute biological samples, the use of mitochondrial DNA to obtain information from old bones and the application of Y chromosomal polymorphisms in sexual assault cases. Present research focuses on genetic markers for external visible traits and on improving the utility of poor quality samples with degraded DNA. The first valid marker for a physical trait was the sex marker amelogenin, which has been part of short tandem repeat (STR) multiplex PCRs for more than a decade. New markers are being investigated for human pigmentation, predicting hair, iris and skin colour. The relation between genes and pigmentation however is complex, involving numerous genes. Single nucleotide polymorphisms (SNPs) in these genes make promising candidates for predictive markers.

Also stature is a complex genetic trait where study subjects can be phenotyped easily. Linkage analysis has revealed several short stature-quantitative trait loci while polymorphisms in the fibrillin I gene are associated with tall stature.

An indirect approach of physical traits is inferring the geographic origin of an individual from DNA. Here, genome wide SNP panels have been used, correctly discriminating "European", "East Asian" and "Sub-Sahara African" ancestry, or even substructuring Europeans to the level of almost nationality, by typing 500 000 SNPs. The ultimate goal is to arrive at “molecular photofitting”, constructing a portrait of the person leaving a biological stain with molecular techniques. The other direction of forensic DNA research addresses the difficulty of obtaining a DNA profile in a degraded crime stain.

One strategy to tackle this problem uses SNPs in stead of STRs. Due to the intrinsic small size of SNPs, they are well suited for degraded samples. There are two major drawbacks however. First, due to their bi-allelic nature, the resolution of a SNP is inferior to any STR. This can be overcome by using larger numbers of SNPs, yielding the same statistical power as 10 to15 STRs. Then, since currently DNA profiles of a standard set of only STRs are recorded in national forensic databases, a DNA profile consisting of only SNPs cannot be matched. An alternative strategy still applies STRs, but with smaller amplicons. Primer pairs are chosen closer to the repeat region, yielding shorter PCR products than with classical primers. The STRs in question are partially “old”, internationally used loci and for another part newly developed systems. Thus, comparison with DNA profiles in existing forensic databases remains possible.

In: Genetics Research Summaries
Editors: L. E. Hewitt and J. W. O'Neil

ISBN: 978-1-62257-646-3

Chapter 75

Application of PCR-DGGE Fingerprinting in Molecular Ecology of Marine Microbial Symbionts

Zhiyong Li*
Marine Biotechnology Laboratory, School of Life Sciences and Biotechnology and Key Laboratory of Microbial Metabolism, Ministry of Education, Shanghai Jiao Tong University, Shanghai, P.R. China

Research Summary

It is now well recognised that only a small fraction of microorganisms can be isolated and characterized, especially marine microbial symbionts because of the difficulty in cultivation. The application of molecular biological methods to study the diversity and ecology of microorganisms in natural environments has been practiced since the mid-1980s. Molecular biological techniques such as genetic fingerprinting techniques provide a pattern or profile of the genetic diversity in a microbial community, and are now frequently used in molecular ecology to explore the microbial diversity and to analyse the structure of microbial communities. Denaturing gradient gel electrophoresis (DGGE) of PCR amplified ribosomal DNA fragments has been introduced into microbial ecology in 1993. Combined with PCR amplification of marker genes or their transcripts, DGGE can give a direct display of the predominant constituents in microbial assemblages with the advantage of easiness, reproducibility, reliability, and speed. PCR-DGGE fingerprinting is suitable for revealing the diversity of uncultured marine microbial symbionts, comparing and monitoring the dynamic change of the community structure of marine microbial symbionts. In this chapter, the theoretical aspects of PCR-DGGE fingerprinting and its application in the molecular ecology of microbial symbionts of marine organisms such as sponge, coral, ascidians, bryozoans and algae are introduced.

* Corresponding author: E-mail: zyli@sjtu.edu.cn; Tel.: (+86)21-34204036.

In: Genetics Research Summaries
Editors: L. E. Hewitt and J. W. O'Neil

ISBN: 978-1-62257-646-3

Chapter 76

Transgene Silencing in Plants: Mechanisms, Applications and New Perspectives

***Chiara Pagliarani*[1]*, Irene Perrone*[1]*, Andrea Carra*[2]
and Giorgio Gambino[3,*]**

[1]Department of Arboriculture and Pomology, University of Torino.
Grugliasco (TO), Italy
[2]Research Unit for Wood Production Outside Forests, CRA,
Casale Monferrato (AL), Italy
[3]Plant Virology Institute, National Research Council, Grugliasco Unit,
Grugliasco (TO), Italy

Research Summary

This review aims to describe the state of and progress in the knowledge of RNA silencing in transgenic plants, including the experimental applications and new perspectives opened by the most recent studies. Modern plant breeding involves new technical approaches, and genetic transformation is undoubtedly a powerful tool in plant biology and plant pathology. However, genetic engineering does not always result in efficient transgene expression, and often transgene copy number does not correlate with transgene expression level. Research in the past decade has shed light on the importance of RNA silencing as a mechanism of virus resistance in transgenic plants. Several plants that are resistant to viruses have been obtained, and some have commercially been applied for crop protection on fields. Transgene silencing is part of a broad host defence system called RNA silencing, a process that leads to homologous RNA degradation, which has widely been observed in animals, plants, and fungi. A key feature of RNA silencing is the presence of small RNAs, such as microRNAs

[*] Author for correspondence. Phone: +39-011-6708666; Fax: +39-011-6708658; E-mail: g.gambino@ ivv.cnr.it

(miRNAs) and small interfering RNAs (siRNAs), which are processed by a member of the RNAse III-like enzyme family, known as DICER. In plants, several distinct RNA silencing pathways operate to repress gene expression at transcriptional or post-transcriptional level. Transcriptional silencing is associated with DNA methylation, in which DNA homologous to a dsRNA is methylated *de novo*. In addition to defence responses against viruses and transposons, short RNAs have been demonstrated to have a role in a diverse range of functions, including regulation of gene expression, development and chromatin structure. RNA silencing is also a powerful tool for functional genomic studies in several species. Transgene-mediated gene silencing through tissue-specific, partial and/or total gene inactivation is a convenient approach to study target genes functions, particularly in species for which mutant collections are not available. We review various strategies for small RNA-based gene silencing: viral expression vectors (virus-induced gene silencing, VIGS), transgenes containing hairpin RNA structures and a recently introduced approach, based on artificial microRNAs (amiRNAs).

In: Genetics Research Summaries
Editors: L. E. Hewitt and J. W. O'Neil
ISBN: 978-1-62257-646-3

Chapter 77

RNA Silencing in the Ectomycorrhizal Fungus *Laccaria Bicolor*

Minna J. Kemppainen and Alejandro G. Pardo*

Laboratorio de Micología Molecular, Departamento de Ciencia y Tecnología, Universidad Nacional de Quilmes, and Consejo Nacional de Investigaciones Científicas y Tecnicas (CONICET).Bernal, Provincia de Buenos Aires, Argentina

Research Summary

Mycorrhiza is a mutualistic association between fungi and the roots of the vast majority of terrestrial plants. In natural ecosystems the plant nutrient uptake from the soil takes place via the extraradical mycelia of these mycosimbionts. While most herbaceous plants and tropical trees form endomycorrhiza-type interactions, trees of boreal and temperate ecosystems are typically ectomycorrhizal (ECM). These species include the majority of ecologically and economically important trees and the fungal symbionts are predominantly filamentous basidiomycetes.

The symbiotic phase in the life cycle of ECM basidiomycetes is the dikaryon. Hence, studies on symbiotic relevant gene functions would require the inactivation of both gene copies in the dikaryotic mycelium.

RNA silencing is a sequence homology-dependent degradation of target mRNAs based on an ancient cellular mechanism believed to have evolved as protection of eukaryotic cells against alien nucleic acids. In different eukaryotic organisms, including fungi, the RNA silencing pathway can be artificially triggered to target and degrade gene transcripts of interest, resulting in gene knock-down. Most importantly, RNA silencing can act at the cytosolic level affecting mRNAs originating from several gene copies and different nuclei, and it can thus offer an efficient way for altering gene expression in dikaryotic organisms.

Laccaria bicolor, the first symbiotic fungus with its genome sequenced, has rapidly turned into a model fungus in ectomycorrhizal research. *Laccaria* possesses a complete set of

* Corresponding author. Phone: 54-11-4365-7100 (ext 4205); Fax: 54-11-4365-7132; E-mail: apardo@unq.edu.ar

genes known to be needed for RNA silencing in eukaryotic cells. We have demonstrated that RNA silencing is functional in *L. bicolor* and that it can be triggered via *Agrobacterium*-mediated transformation. Moreover, targeted gene knock-down in dikaryotic mycelium can result in functional phenotypes altered in the symbiotic capacity confirming that RNA silencing is a powerful way to study symbiosis- regulated genes. These findings have now initiated the RNA silencing era in mycorrhizal research, a field that has been hindered by the lack of proper genetic tools.

In: Genetics Research Summaries
Editors: L. E. Hewitt and J. W. O'Neil

ISBN: 978-1-62257-646-3

Chapter 78

Gene Silencing as a Promising Strategy to Target CNS and Neurological Disorders

***Ana L. Cardoso*[1], *Sara M. Trabulo*[1,2], *Sónia P. Duarte*[1,2], *Pedro M. Costa*[1,2], *Luís P. Almeida*[1,3] *and Maria C. Pedroso de Lima*[1,2]**

[1]Center for Neuroscience and Cell Biology, University of Coimbra, Coimbra, Portugal
[2]Department of Life Sciences, Faculty of Science and Technology, University of Coimbra, Coimbra, Portugal
[3]Laboratory of Pharmaceutical Technology, Faculty of Pharmacy, University of Coimbra, Coimbra, Portugal

Research Summary

RNA interference (RNAi) has recently emerged as a powerful tool in functional genomic studies, allowing dissection of entire signalling pathways and elucidation of the molecular mechanisms of neurobiological processes, thereby facilitating rapid identification and validation of possible therapeutic targets. Moreover, RNAi holds great therapeutic potential since application of small interfering RNAs (siRNAs) and short hairpin RNAs (shRNAs) may allow specific knockdown of selected toxic proteins, even when allele-specific silencing is needed, as in the case of dominantly inherited disorders.

Nevertheless, the development of RNAi-based therapeutics for *in vivo* application faces the same challenge common to all classes of drugs: achieving an efficient and sustained distribution into the target tissue at sufficient concentrations to accomplish a therapeutic effect. Although significant progress has been made regarding the safety and stability of siRNAs and shRNAs, a major limitation for the *in vivo* application of RNAi technology concerns the inability of these molecules to cross cellular membranes. Multiple delivery methodologies, including viral and non-viral vectors, have been developed with different degrees of success for the introduction of siRNAs and shRNAs into cells, both *in vitro* and *in vivo*.

This review is focused on the available strategies to achieve gene silencing in the CNS and on the most extensively studied systems to mediate siRNA and shRNA delivery into the brain. Moreover, we summarize the most important studies concerning RNAi application in the context of neurodegenerative diseases and other neurological disorders.

In: Genetics Research Summaries
Editors: L. E. Hewitt and J. W. O'Neil

ISBN: 978-1-62257-646-3

Chapter 79

Polymeric Nanoparticles as Efficient siRNA Delivery Systems

***Surendra Nimesh*[1,*], *Nidhi Gupta*[1] *and Ramesh Chandra*[2,3]**
[1]Laboratory of Biochemical Neuroendocrinology,
Clinical Research Institute of Montreal, Montreal, Quebec, Canada
[2]Dr. B. R. Ambedkar Center for Biomedical Research,
University of Delhi, Delhi, India
[3]Department of Chemistry, University of Delhi, Delhi, India

Research Summary

Small interfering RNAs (siRNA) are emerging as promising therapeutic agents for the treatment of inherited and acquired diseases, as well as research tools for the elucidation of gene function. Since the molecules undergo rapid enzymatic degradation and have poor cellular uptake, there is a need to design a delivery system which can protect and efficiently transport siRNA to the target cells. Polymeric nanoparticles have emerged as systems of choice with reduced cytotoxicity and enhanced efficacy. These systems not only protect siRNA from enzymatic degradation by forming condensed complexes but also leads to tissue and cellular targeting, improve cellular penetration, release the siRNA in the right intracellular compartment. Nanoparticles prepared from polycationic polymers like polyethylenimine, chitosan have been widely investigated due to ease of manipulatibility, stability, low immunogenicity, low cost and high flexibility regarding the size of transgene delivered. This chapter presents an overview of siRNA delivery strategies employing polymeric nanoparticles, with emphasis on self-assembled polymeric nanoparticles with promising potential to evolve as therapeutic tool in gene therapy.

* Corresponding author. Laboratory of Biochemical Neuroendocrinology, Clinical Research Institute of Montreal, 110 West Pine Ave, Montreal, Quebec, Canada H2W 1R7. Tel: 514-987-5559. Fax: 514-987-5542. E-mail address: surendranimesh@gmail.com

In: Genetics Research Summaries
Editors: L. E. Hewitt and J. W. O'Neil
ISBN: 978-1-62257-646-3

Chapter 80

Application of Negatively Charged Hydroxyproline-based DNA Mimics for Gene Silencing

Vladimir A. Efimov and Oksana G. Chakhmakhcheva
Shemyakin-Ovchinnikov Institute of Bioorganic Chemistry,
Moscow, Russia

Research Summary

With the aim in view to improve physicochemical and biological properties of natural oligonucleotides, several types of DNA analogues and mimics were designed, particularly negatively charged PNA-like mimics. Among them, two types of DNA mimics representing hetero-oligomers constructed from alternating monomers of phosphono peptide nucleic acids and monomers on the base of *trans*-1-acetyl-4-hydroxy-L-proline (HypNA-pPNAs) as well as oligomers constructed from chiral analogues of peptide nucleic acids with a constrained trans-4-hydroxy-N-acetylpyrrolidine-2-phosphonic acid backbone (pHypNAs) were developed. Their physico-chemical and biological properties were evaluated in the comparison with natural oligonucleotides, classical peptide nucleic acids and morpholino phosphorodiamidate oligonucleotide analogues. The results obtained in a set of experiments revealed a high potential of these phosphonate-containing PNA derivatives for a number of biological applications, such as diagnostic, nucleic acids analysis and inhibition of gene expression. HypNA-pPNA and pHypNA mimics combine high hybridization and mismatch discrimination characteristics with good water solubility and biological stability as well as the ability to penetrate cell membranes. Their effectiveness to provide the specific knockdown of a target protein production was demonstrated in research involving *in vitro* systems, living cells and intact organisms. As their effect lasts over a period of several days, due to their high stability in living cells, it represents a very potent technology for administrating antisense- or antigene-based drugs for future therapeutic applications.

In: Genetics Research Summaries
Editors: L. E. Hewitt and J. W. O'Neil

ISBN: 978-1-62257-646-3

Chapter 81

Gene Silencing and the Analysis of Immune Response in Model Insects

Davide Malagoli and Mauro Mandrioli*
Department of Biology, University of Modena and Reggio Emilia,
Modena, Italy

Research Summary

Insects are organisms of considerable interest for comparative biology and medicine, therefore it is not surprising that several publications referred to them as model organisms. Insect and vertebrate evolution diverged more than 500 million years ago, but the molecular bases of several fundamental biological functions, including innate immune response, were already established in their common progenitor and have been conserved. Consequently, starting from information collected in insects, new insights into human biology and pathology were gained. Gene silencing includes several powerful methods, such as the production of loss-of-function mutants and RNA interference. These procedures, in particularly when performed in models for which molecular databases are already available, allow the genetic dissection of several immune-related processes and pathways. In the present review, we will concentrate our attention on the information derived from gene silencing techniques on insect immune signalling with particular attention for *Drosophila melanogaster* and *Anopheles gambiae*.

* Correspondence to: Mauro Mandrioli, Department of Biology, University of Modena and Reggio Emilia, Modena, Via Campi 213/D, 41125, Italy, E-mail: mauro.mandrioli@unimore.it

In: Genetics Research Summaries
Editors: L. E. Hewitt and J. W. O'Neil

ISBN: 978-1-62257-646-3

Chapter 82

RNAi in Plants: Recent Developments and Applications in Agriculture

Muthappa Senthil-Kumar and Kirankumar S. Mysore*
The Samuel Roberts Noble Foundation, Ardmore, OK, US

Research Summary

Sequencing of plant genome and expressed sequence tag (EST) have provided abundant sequence information in several plant species. Elucidating function(s) of all of these genes is a huge undertaking. Even in well-studied plants like Arabidopsis, function is not known for majority of genes. Hence, a powerful tool that can be widely used to understand gene function is necessary. Several functional genomics tools were developed in the recent past to achieve this goal. RNA interference (RNAi) is one such tool widely used to analyze gene function. RNAi is also proved to be a tool for plant researchers to produce improved crop varieties. First part of this review is focused on three RNAi based concepts that has potential applications in plant functional genomics and agriculture. These concepts are tissue specific silencing, inducible silencing and host delivered RNAi (hdRNAi) during plant-pest interaction.

Tissue specific promoters driving RNAi constructs can induce gene silencing in a particular organelle or tissue. Also, RNAi constructs with stress or chemical inducible promoters can be used to induce gene silencing only when required. These two concepts together can be used to achieve temporal and spatial control of gene silencing in plants. In the hdRNAi, dsRNA generated in an RNAi transgenic plant is delivered to interacting target organism (pest), activating gene silencing in the target organism. A comprehensive review pertaining to these areas is presented. Second part of the review deals with applications of RNAi in agriculture, animal husbandry and biofuel industry. As suppression of gene expression by RNAi is inheritable, this has been a tool for developing transgenic crop plants

* Corresponding author. Plant Biology Division, The Samuel Roberts Noble Foundation, 2510 Sam Noble Pky, Ardmore, OK 73402 USA. Email: ksmysore@noble.org Fax: 580-224-6692

for resistance against disease, pests, drought and in other areas of agriculture. This review summarizes developments in these areas with major emphasis on application of RNAi for development of biotic stress tolerant crops. We also note limitations of RNAi technology and ways to overcome the same.

In: Genetics Research Summaries
Editors: L. E. Hewitt and J. W. O'Neil

ISBN: 978-1-62257-646-3

Chapter 83

Silencing of Galectin-3 by Promoter Methylation during Prostate Cancer Progression: A Novel Transcriptional Regulation of Galectin-3 Expression

Hafiz Ahmed[*]

Department of Biochemistry and Molecular Biology, Program in Oncology, Greenebaum Cancer Center, University of Maryland School of Medicine, Baltimore, MD, US

Research Summary

Protein-carbohydrate interactions play significant role in modulating cell-cell and cell-extracellular matrix interactions, which, in turn, mediate various biological processes such as growth regulation, immune function, cancer metastasis, and apoptosis. Galectin-3, a member of the β-galactoside-binding protein family, is found multifunctional and is involved in normal growth development as well as cancer progression and metastasis, but the detailed mechanisms of its functions or its transcriptional regulations are not well understood. Besides, several regulatory elements such as GC box, CRE motif, AP-1 site, and NF-κB sites, the promoter of galectin-3 gene (*LGALS3*) contains several CpG islands that can be methylated during tumorigenesis leading to the gene silencing. This review discusses the galectin-3 epigenetics, which represents a novel regulatory mechanism of its transcription.

[*] Address for correspondence: Hafiz Ahmed, Ph.D. Columbus Center. 701 East Pratt Street, Suite 236, Baltimore, MD 21202, USA. E-mail: ahmed@umbi.umd.edu; hahmed@som.umaryland.edu

In: Genetics Research Summaries
Editors: L. E. Hewitt and J. W. O'Neil
ISBN: 978-1-62257-646-3

Chapter 84

Silencing Hepatitis B Virus Replication with Antiviral Pri-Mir Shuttles Generated from Liver-Specific Pol II Promoters

Abdullah Ely and Patrick Arbuthnot*
Antiviral Gene Therapy Research Unit,
Department of Molecular Medicine and Haematology,
University of the Witwatersrand Medical School, WITS, South Africa

Research Summary

Chronic infection with hepatitis B virus (HBV) occurs in approximately 6% of the world's population and is often complicated by cirrhosis and hepatocellular carcinoma (HCC). Existing therapy rarely has durable effects and improving treatment to counter the infection remains an important medical priority. Although harnessing the RNA interference (RNAi) pathway to achieve therapeutic HBV gene silencing holds promise, precise regulation of the expression of silencing sequences is critically important for safe application of this approach. Earlier work from our laboratory demonstrated that pri-miR-31- and pri-miR-122-based anti HBV shuttles were capable of potent, safe antiviral activity and can be used in modular multimeric arrangements. To advance this approach, and limit the potential problems caused by extrahepatic expression of anti HBV RNAi activators, these sequences were placed under control of liver specific transcription control elements, viz. the human Factor VIII (FVIII), alpha-1-antitrypsin (A1AT), HBV preS2 and HBV basic core (BCP) promoters. Using a luciferase reporter gene assay optimal liver-specific transcription control was observed with A1AT and BCP regulating sequences. These elements were then incorporated into pri-miR-expression cassettes and were tested for antiviral efficacy in cell culture and a

* Corresponding author: Tel: +27 (0)11 717 2365 ; Fax: +27 (0)11 717 2395 ; E-mail: Patrick.Arbuthnot@wits.ac.za

murine model of HBV replication. Results showed that silencing of HBV replication was achieved. Importantly there was no evidence for disruption of endogenous miR function, which is a significant advantage over use of stronger and constitutively active RNA polymerase (Pol) III promoter RNAi expression cassettes. The use of anti HBV pri-miR shuttles in the context of liver-specific Pol II promoters is likely to have usefulness for therapeutic HBV knockdown, and should also have general applicability to silencing of pathology causing genes in the liver.

In: Genetics Research Summaries
Editors: L. E. Hewitt and J. W. O'Neil

ISBN: 978-1-62257-646-3

Chapter 85

Effective Methods for Selecting siRNA Sequences by Using the Average Silencing Probability and a Hidden Markov Model

Shigeru Takasaki*
Toyo University, Ora-gun Gunma, Japan

Research Summary

Short interfering RNA (siRNA) has been widely used for studying gene functions in mammalian cells but varies markedly in its gene silencing efficacy. Although many design rules/guidelines for effective siRNAs based on various criteria have been reported recently, there are only a few consistencies among them. This makes it difficult to select effective siRNA sequences in mammalian genes. This chapter first reviews the recently reported siRNA design guidelines and then proposes a new method for selecting effective siRNA sequences from many possible candidates by using the average silencing probability on the basis of a large number of known effective siRNAs. It is different from the previous score-based siRNA design techniques and can predict the probability that a candidate siRNA sequence will be effective. The results of evaluating it by applying it to recently reported effective and ineffective siRNA sequences for various genes indicate that it would be useful for many other genes. The evaluation results indicate that the proposed method would be useful for many other genes. It should therefore be useful for selecting siRNA sequences effective for mammalian genes. The chapter also describes another method using a hidden Markov Model (HMM) to select the optimal functional siRNAs and discusses the frequencies

* Correspondence should be addressed to e-mail: s_takasaki @ toyonet.toyo. ac.jp Phone: +81-276-82-9024, Fax: +81-276-82-9033

of the combinations for two successive nucleotides as important characteristics of effective siRNA sequences.

In: Genetics Research Summaries
Editors: L. E. Hewitt and J. W. O'Neil

ISBN: 978-1-62257-646-3

Chapter 86

Small RNA-mediated Gene Silencing for Plant Biotechnology

Ulku Baykal and Zhanyuan Zhang*
Plant Transformation Core Facility, Division of Plant Sciences,
University of Missouri, Columbia, MO, US

Research Summary

Small RNA-mediated gene silencing as a natural defense mechanism against viruses, transposons, and other invading nucleic acids or a means of regulating plant endogenous genes is a powerful tool and is being employed to down-regulate the expression of the targeted genes. Such a small RNA-mediated gene silencing has many different applications in a variety of organisms including humans and animals to control disease as a therapeutic agent, as well as plants to alter plant phenotypes. This silencing platform works through RNA-directed degradation or translational repression of target mRNA and has been devised towards a high-throughput approach for the gene suppression. In particular, sequence-specific control of gene expression by these non-coding RNAs has gained a significant amount of importance in plant biotechnology to influence specific plant phenotypes over the past years. It has been demonstrated that crops that were transformed with RNAi constructs, introduced stable modifications to the biochemical pathways. This can open new avenues in the improvement of crop productivity and quality. Here, we review the role of small RNA-directed gene silencing in plant biotechnology. The review will focus on the application of a gene silencing approach mediated by three subclasses of small RNAs for improved oil quality, reduced allergen, virus resistance, and other agronomical traits. The advantages and drawbacks of each gene silencing approach are also discussed with regard to crop improvement.

* Correspondence: 1-31 Agriculture Building, University of Missouri, Columbia, MO 65211, USA. Tel: 573-882-6922; Email: zhangzh@missouri.edu

In: Genetics Research Summaries
Editors: L. E. Hewitt and J. W. O'Neil
ISBN: 978-1-62257-646-3

Chapter 87

Post Transcriptional Gene Silencing Methods for Functional Characterization of Abiotic Stress Responsive Genes in Plants

Muthappa Senthil-Kumar [*1, 2] ***and Makarla Udayakumar*** [1]

[1]Department of Crop Physiology, University of Agricultural Sciences, GKVK, Bangalore, India

[2]Plant Biology Division, The Samuel Roberts Noble Foundation, Ardmore, OK, US

Research Summary

Several abiotic stress specific functional and regulatory genes have been cloned, and a number of EST databases representing stress specific genes are available for many plant species. These sequences have to be translated into functional information, necessitating the need for potential functional genomic approaches. Post transcriptional gene silencing (PTGS) is one of approaches to characterize functional relevance of stress responsive genes. Virus-induced gene silencing (VIGS) and developing stable gene knock down plants using hairpin RNA interference (hpRNAi) constructs (referred here as RNAi) are two important PTGS methods. Over a period, these methods are becoming integral part of plant stress functional genomics. Among these two methods, use of VIGS for characterizing abiotic stress responsive genes is still an emerging approach while RNAi has been widely used.

This review is focused on VIGS vector resources, brief methodology of VIGS and application of gene silencing to identify/characterize genes involved in drought-, salinity-oxidative-, high light-, and nutrient-stress management. VIGS can be used as fast forward

[*] Corresponding author: Muthappa Senthil-Kumar, Plant Biology Division, Noble Foundation, 2510 Sam Noble Pky, Ardmore, OK 73402 USA, Email: skmuthappa@noble.org , or senthilphy@yahoo.com.

genetic screening method to identify genes involved in stress tolerance and also an effective reverse genetic tool to validate the relevance of genes identified from high-throughput screening. Further, VIGS can be effectively integrated with abiotic stress imposition and response of gene silenced plants can be quantified using suitable techniques. We describe here an comprehensive approach to silence large number of cDNA clones and characterize the silenced plants under abiotic stresses. We also discussed application of other PTGS based methods like RNAi and artificial micro RNA (amiRNA) in abiotic stress functional genomics.

We propose that PTGS is a useful technology for translational genomics to assign function to large number of abiotic stress responsive genes. Even with their current limitations, gene silencing techniques are set to revolutionize plant abiotic stress functional genomics. Limitations and future directions for these techniques are also briefly discussed.

In: Genetics Research Summaries
Editors: L. E. Hewitt and J. W. O'Neil

ISBN: 978-1-62257-646-3

Chapter 88

Genome-wide Identification and Analysis of miRNAs Complementary to Upstream Sequences of mRNA Transcription Start Sites

Kenta Narikawa[1], *Kenji Nishi*[2], *Yuki Naito*[2], *Minami Mazda*[2] *and Kumiko Ui-Tei1*[*,2]
[1]Department of Computational Biology, Graduate School of Frontier Sciences, University of Tokyo, Kashiwa-shi, Japan
[2]Department of Biophysics and Biochemistry, Graduate School of Science, University of Tokyo, Bunkyo-ku, Tokyo, Japan

Research Summary

Small regulatory RNAs including short interfering RNAs (siRNAs) and microRNAs (miRNAs) are crucial regulators of gene expression at the posttranscriptional level. Recently, additional roles for small RNAs in gene activation and suppression at the transcriptional level were reported; these RNAs were shown to have sequences that closely or completely match to their respective promoter regions. However, no global analysis for identifying target sequences for miRNAs in the promoter region have been carried out in the human genome.

We performed a genome-wide search for upstream sequences of mRNA transcription start sites where miRNAs are capable of hybridizing with high complementarity. We identified 219 sites in the 10-kb upstream regions of transcription start sites with complete complementarity to 94 human mature miRNAs. Furthermore, the mismatched positions and nucleotides in near-completely matched sites were highly biased, and most of them appear to be possible target sites of miRNAs. The expression of downstream genes of miRNA target sites were examined following transfection of each miRNA into three different human cell

* Email address: ktei@bi.s.u-tokyo.ac.jp

lines. The results indicate that miRNAs dynamically modulates gene expression depending on the downstream genes and the cell type.

In: Genetics Research Summaries
Editors: L. E. Hewitt and J. W. O'Neil
ISBN: 978-1-62257-646-3

Chapter 89

Gene Silencing: Theory, Techniques and Applications

Prashant Mohanpuria, Vinay Kumar, Monika Mahajan, Hasan Mohammad and Sudesh Kumar Yadav [*]
Biotechnology Division, Institute of Himalayan Bioresource Technology, CSIR, Palampur, India

Research Summary

Gene silencing is an exciting field of functional genomics. It has been used as a research tool to discover or validate the functions of genes. It involves short sequence of nucleic acid that can bind to RNA of the gene and interferes the process of its expression. It is diverse in occurrence as well as in applications. This phenomenon occurs from nematodes to fungi and can cause gene silencing in plants, animals and human beings. The core aspects of the mechanisms and functions of gene silencing include co-suppression, RNA-mediated virus resistance and RNA-directed DNA methylation (RdDM). The applications of gene silencing cover a wide spectrum in plants, from designer flower colors to plant-produced medical therapeutics. These functions are achieved by two types of approach such as protection of the plant against attack and fine-tuning of metabolic pathways. RNA-mediated gene control mechanism has already provided new platforms for developing molecular tools for gene function studies and crop improvements. We are now exploring this technology for commercially focused applications in plants. Here, we review the theory of gene silencing discovery and the mechanism of this technique in plants. Further, we discuss the potential use of this technique in plant science particularly in crop improvements.

[*] Corresponding author: skyt@rediffmail.com; sudeshkumar@ihbt.res.in

In: Genetics Research Summaries
Editors: L. E. Hewitt and J. W. O'Neil

ISBN: 978-1-62257-646-3

Chapter 90

RNA Interference Mediated by Cell-Penetrating Peptides

Cheng-Yi Lee, Betty Revon Liu, Ya-Hui Wang, Yu-Wun Hou, Jyh-Ching Chou and Han-Jung Lee*

Department of Natural Resources and Environmental Studies,
National Dong Hwa University, Hualien, Taiwan

Research Summary

RNA interference (RNAi) has been utilized in a variety of applications to target specific gene silencing mediated by small-interfering RNA (siRNA) over the last few years. Cell-penetrating peptides (CPPs) were proven to be able to traverse cell membranes and deliver biological macromolecules into living cells. Here, we provided an efficient and safe method for the delivery of siRNA into mammalian cells mediated by CPPs noncovalently. We first established a GC-EGFP cell line stably expressing enhanced green fluorescent protein (EGFP) from human gastric cells. CPPs were demonstrated to interact with and deliver siRNA into GC-EGFP cells, and the internalized dsRNA tended to localize in the perinuclear region within cells. The sulforhodamine B (SRB) assay further confirmed CPPs were nontoxic to cell viability. Finally, our results showed that siRNA fulfilled its targeted *egfp* gene silencing. In the future, CPPs may provide a useful and nontoxic tool for the delivery of siRNA into mammalian cells.

* Corresponding author: E-mail: hjlee@mail.ndhu.edu.tw

In: Genetics Research Summaries
Editors: L. E. Hewitt and J. W. O'Neil
ISBN: 978-1-62257-646-3

Chapter 91

Asymmetric RNA Duplexes as Next Generation RNAi Inducers

***Dong-ki Lee*[*1], *Soyoun Kim*[2], *Andrew C. Keates*[3] *and Chiang J. Li*[2,3]**

[1]Global Research Laboratory for RNAi Medicine, Department of Chemistry and BK21 School of Chemical Materials Science, Sungkyunkwan University, Suwon, Korea
[2]Department of Biomedical Engineering, Dongguk University, Seoul, Korea
[3]Skip Ackerman Center for Molecular Therapeutics, Beth Israel Deaconess Medical Center, Harvard Medical School, Boston, MA, US

Research Summary

RNA interference (RNAi) has become an indispensable technology for biomedical research and promises to usher in a brand new class of therapeutics that work by silencing disease genes. Until recently, the paradigm for gene silencing in mammalian cells has relied on a small symmetrical RNA structures containing a 19-base-pair duplex with 2 nucleotide overhangs at each 3' end: the standard siRNA structure. The standard siRNA scaffold is based on structures generated by Dicer digestion of a double stranded RNA, and is considered to be the fundamental template for designing RNAi inducers. In fact, early studies suggested there was only very limited flexibility regarding the length and symmetry of the siRNA structure in order to maintain optimal gene silencing. Recent studies, however, have demonstrated that gene silencing siRNAs with duplex lengths shorter than 19 bp or asymmetric structures can trigger specific gene silencing in mammalian cells. Importantly, asymmetric siRNA structures can ameliorate several sequence-independent, nonspecific effects triggered by the canonical siRNA structure. These findings demonstrate the structural flexibility of RNAi inducers in mammalian cells.

[*] Correspondence: Dong-ki Lee, dklee@skku.edu † Chiang J. Li, cli@bidmc.harvard.edu

In: Genetics Research Summaries
Editors: L. E. Hewitt and J. W. O'Neil
ISBN: 978-1-62257-646-3

Chapter 92

Analysis of Sequence Diversity at Two Mitochondrial Genes on Different Taxonomic Levels. Applicability of DNA Based Distance Data in Genetics of Speciation and Phylogenetics

Y. Ph. Kartavtsev
A. V. Zhirmunsky Institute of Marine Biology of the Far Eastern Branch of the Russian Academy of Sciences, Vladivostok, Russia

Research Summary

Algorithms of nucleotide diversity estimates and other measures of genetic divergence for the two genes *Cyt-b* (cytochrome *b*) and *Co-1* (cytochrome oxidase 1) are analyzed. Based on the theory and algorithms of distance estimates on DNA sequences, as well as on the observed distance values retrieved from literature, it is recommended for realistic tree building to use a specific nucleotide substitution model from at least 56 available from Modeltest 3.7 or other software depending on the specific set of nucleotide sequences. Using a database of *p*-distances and similar measures gathered from published sources and GenBank (http://www.ncbi.nlm.nih.gov) sequences, genetic divergence of populations (1) and taxa of different rank, such as subspecies, semispecies or/and sibling species (2), species within a genus (3), species from different genera within a family (4), and species from separate families within an order (5) have been compared.

Empirical data for 18,192 vertebrate and invertebrate species demonstrate that the data series are realistic and interpretable when *p*-distance and its various derivates are used. The focus was on vertebrates and fish species in particular, and the newest dataset obtained in the framework of FishBOL (http://www.fishbol.org). Distance data revealed various and increasing levels of genetic divergence of the sequences of the two genes *Cyt-b* and *Co-1* in the five groups compared. Mean unweighted scores of *p*-distances for five groups are: *Cyt-b* (1) 1.46±0.34, (2) 5.35±0.95, (3) 10.46±0.96, (4) 17.99±1.33 (5) 26.36±3.88 and *Co-1* (1) 0.72±0.16, (2) 3.78±1.18, (3) 10.87±0.66, (4) 15.00±0.90, (5) 19.97±0.80. The estimates show good correspondence with former analyses. This testifies to the applicability of *p*-distance for most intraspecies and interspecies comparisons of genetic divergence up to the order level for the two genes compared. As seen from the numbers above, and from a regression analysis, there is no a sign of saturation, usually expected from a homoplasy effect.

Differences in divergence between the genes themselves at the five hierarchical levels were also found. This conforms to the ample evidence showing different and nonuniform evolution rates of these and other genes and their various regions. The results of the analysis of the nucleotide as well as allozyme divergence within species and higher taxa of animals are, firstly, in a good agreement with previous results and showed the stability of a general trend, and, secondly, suggest that in animals, phyletic evolution is likely to prevail at the molecular level, and speciation mainly corresponds to the geographic model (type D1). The prevalence of the D1 speciation mode does not mean that other modes are absent. There are at least seven possible modes of speciation. How we can recognize them formally with operational genetic criteria is a key question for establishing a quantifiable genetic model (theory) of speciation. An approach is suggested that allows a step forward in this direction.

Research was supported by the Russian Foundation for Fundamental Research grants *#07-04-00186, #08-04-91200 and the Far Eastern Branch of the Russian Academy of Sciences (RAS) grant #08-3B-06-031, RAS Board Programs, grant #09-1P23-06.*

In: Genetics Research Summaries
Editors: L. E. Hewitt and J. W. O'Neil
ISBN: 978-1-62257-646-3

Chapter 93

Chromosomal Variability and the Origin of *Citrus* Species

Marcelo Guerra
Department of Botany, Federal University of Pernambuco
Recife, PE, Brazil

Research Summary

The genus *Citrus* includes some of the most important crop plants in the world although its taxonomy remains one of the most controversial among angiosperms. Most species are of hybrid origin and some of them may include germplasm from other genera. Cytologically, *Citrus* species are characterized by a stable chromosome number and a highly variable pattern of heterochromatic bands. Most accessions display heteromorphic chromosome pairs, suggesting that they were originated from cross hybridization. On the other hand, citron (*C. medica*), pummelo (*C. maxima*), a few mandarin accessions, and most wild *Citrus* species and related genera exhibit chromosome pairs that are homomorphic for similar heterochromatic bands. Based on these findings, hybrids and non-hybrid accessions were identified and the possible origin and relationship among most accessions were reconsidered.

In: Genetics Research Summaries
Editors: L. E. Hewitt and J. W. O'Neil
ISBN: 978-1-62257-646-3

Chapter 94

Genetic Diversity of *Mycobacterium Tuberculosis* Population in Bulgaria

***Violeta Valcheva*[1], *Igor Mokrousov*[2 *], *Olga Narvskaya*[2], *Nalin Rastogi*[3] *and Nadya Markova*[1]**

[1]Department of Pathogenic Bacteria, The Stephan Angeloff Institute of Microbiology, Bulgarian Academy of Sciences, Sofia, Bulgaria
[2]Laboratory of Molecular Microbiology, St. Petersburg Pasteur Institute, St. Petersburg, Russia
[3]Unité de la Tuberculose et des Mycobactéries, Institut Pasteur de Guadeloupe, Abymes, Guadeloupe

Research Summary

Tuberculosis remains an important public health issue for Bulgaria, a Balkan country located in the world region with contrasting epidemiological situation for tuberculosis. Here, we present results of the recent studies on the genetic diversity of *Mycobacterium tuberculosis* population in Bulgaria that was evaluated with various DNA fingerprinting methods (spoligotyping, 24-MIRU-VNTR and IS*6110*-RFLP typing). The spoligotype-based population structure of *M. tuberculosis* in Bulgaria was shown to be sufficiently heterogeneous. It is dominated by several worldwide distributed spoligotypes ST53 and ST47 and Balkan-specific spoligotypes ST125 and ST41. The Beijing genotype strains were not found in Bulgaria in spite of close links with Russia in the recent and historical past. Comparison with international database SITVIT2 (Pasteur Institute of Guadeloupe) showed that spoligotype ST53 is found in similar and rather high proportion in the neighboring Greece and Turkey and almost equally distributed across different regions of Bulgaria. Contrarily, ST125 is not found elsewhere and is specific for Bulgaria; furthermore it appears

* Correspondence: Igor Mokrousov, Ph.D., St. Petersburg Pasteur Institute, 14 Mira street, St. Petersburg, 197101 Russia. Email: imokrousov@mail.ru

to be mainly confined to the southern part of the country. Novel 15/24-loci format of MIRU-VNTR typing was found to be the most discriminatory tool compared to spoligotyping and IS*6110*-RFLP typing of *M. tuberculosis* strains in Bulgaria. Furthermore, VNTR typing was shown useful for resolving ambiguous phylogeny of some spoligotypes, in particular, those classified as LAM/S by bioinformatics approach. In practical terms, a reduced "Bulgaria-specific" 5-locus set (MIRU40, Mtub04, Mtub21, QUB-11b, QUB-26) provided a sufficiently high differentiation and may be preliminarily recommended for a first-line typing of *M. tuberculosis* isolates in Bulgaria although further studies are needed to validate this scheme. At the same time, a comprehensive secondary subtyping of the clustered isolates should target all 15 discriminatory loci. We additionally investigated molecular basis of drug resistance of the studied strains. Three types of the *rpoB* mutations were found in 20 of 27 RIF-resistant isolates; *rpoB* S531L was the most frequent. Eleven (48%) of 23 INH-resistant isolates had *katG* S315T mutation. *inhA* -15C>T mutation was detected in one INH-resistant isolate (that also had *katG315* mutation) and three INH-susceptible isolates. A mutation in *embB306* was found in 7 of 11 EMB-resistant isolates. Consequently, *rpoB* and *embB306* mutations may serve for rapid genotypic detection of the majority of the RIF and EMB-resistant strains in Bulgaria; the results on INH resistance are complex and further investigation of more genes is needed. Comparison with spoligotyping and 24-VNTR locus typing data suggested that emergence and spread of drug-resistant and MDR-TB in Bulgaria are not associated with any specific spoligotype or MIRU-VNTR genotype. A local circulation of the particular clones appears to be an important factor to take into consideration in the molecular epidemiological studies of tuberculosis in Bulgaria.

In: Genetics Research Summaries
Editors: L. E. Hewitt and J. W. O'Neil

ISBN: 978-1-62257-646-3

Chapter 95

Genetic Diversity in Switchgrass – A Potential Bioenergy Crop

B. Narasimhamoorthy, M. C. Saha, H. S. Bhandari and J. H. Bouton*
Forage Improvement Division,
The Samuel Roberts Noble Foundation, Inc.,
Ardmore, OK, US

Research Summary

Switchgrass (*Panicum virgatum* L.) is a warm-season C4 perennial grass belonging to the family Poaceae. It is native to North America. Persistence across a wide geographical range, in addition to high biomass production with minimum inputs, makes it an excellent choice for a sustainable bioenergy crop. Switchgrass is a highly heterozygous, self-incompatible and out-crossing species. Broad species adaptation, natural selection and photoperiodism have combined to create considerable ecotypic differentiation in switchgrass. The natural population is classified into two distinct cytotypes; upland and lowland. Upland cytotypes are mostly octaploid ($2n = 8x = 72$) and lowlands are tetraploid ($2n = 4x = 36$); however, multiple ploidy levels ranging from diploid ($2n = 2x = 18$) to dodecaploid ($2n = 12x = 108$) have been reported in switchgrass. In the USA, uplands are adapted to the mid and northern latitudes, while lowlands are in the southern parts of the country. In addition, these ecotypes differ with respect to photosynthesis, drought tolerance and N-use efficiency. Knowledge on the amount of genetic diversity and polymorphism in switchgrass is necessary to enhance the effectiveness of breeding programs and germplasm conservation efforts. In the past two decades, several studies have been conducted to evaluate the genetic variability in switchgrass populations. Molecular markers, such as RFLPs, RAPDs and SSRs, were used to find within

* Author for correspondance : Malay C. Saha, Forage Improvement Division, The Samuel Roberts Noble Foundation, Inc., 2510 Sam Noble Parkway, Ardmore, OK 73401, USA, E-mail: mcsaha@noble.org, Phone: +1-580-224-6840. Fax: +1-580-224-6802.

and among population variation in a wide range of switchgrass cytotypes. Hybrid cultivars can be an attractive option for improving biomass production. Molecular marker and phenotypic data suggest that lowland and upland genotypes represent different heterotic groups that can potentially be used to produce F_1 hybrid cultivars. This review summarizes the current understandings on the genetic diversity available in *P. virgatum* populations, with a focus on studies performed at the Noble Foundation, where the genetic variability and the relationships within and among switchgrass populations were determined with simple sequence repeat markers and ploidy analysis.

In: Genetics Research Summaries
Editors: L. E. Hewitt and J. W. O'Neil
ISBN: 978-1-62257-646-3

Chapter 96

Genetic Variability in the Fescue-Ryegrass Complex

F. M. Kirigwi, A. A. Hopkins and M. C. Saha*
Forage Improvement Division, The Samuel Roberts Noble Foundation, Inc., Ardmore, OK, US

Research Summary

Fescues and ryegrasses in the *Lolium* genus are widely used as forage and turf, especially in temperate regions of the world. These highly productive grass species provide feed and fodder for livestock and wild animals, play a major role as turf on golf courses and lawns worldwide, and prevent soil erosion. Among these grasses, tall fescue [*Lolium arundinaceum* (Schreb.) Darbysh.] germplasm is classified into five botanical varieties that range from tetraploid to decaploid and into two major germplasm pools, "Continental" and "Mediterranean," as well as into two functional groups, forage and turf types. Important species in the genus *Lolium* include the outcrossing *Lolium perenne* L., (perennial ryegrass) and the self-pollinated *L. temulentum* L. subsp. *temulentum* (darnel, darnel ryegrass). The majority of the *Lolium* are self-infertile, have a strong self-incompatibility system and are, therefore, highly heterogeneous. Grazing or selection may lead to loss of rare alleles that may be useful in adaptation in extreme environments, e.g., when these cool-season grasses are grown in warmer, drier areas. Understanding the levels of genetic diversity within and genetic relationships between populations is therefore important for not only breeding, but also for ensuring adaptability and persistence, quality and disease resistance of germplasm accessions, breeding lines and populations. At the Noble Foundation, efforts have been concentrated on collecting tall fescue and *L. temulentum* germplasm, and the development of molecular tools for these species. Molecular tools developed in-house were employed to study genetic

* Author for correspondence: Malay C. Saha, Forage Improvement Division. The Samuel Roberts Noble Foundation, Inc. 2510 Sam Noble Parkway. Ardmore, OK 73401, USA. E-mail: mcsaha@noble.org. Phone: +1-580-224-6840. Fax: +1-580-224-6802

diversity and to understand the utility of various marker tools for diversity studies. In this chapter, we review the genetic diversity work carried out in *Lolium*, with an emphasis on our work at the Noble Foundation. Various marker systems have been found to be useful in the *Lolium* genus, with SSRs in particular being transferable across the fescue-ryegrass complex.

In: Genetics Research Summaries
Editors: L. E. Hewitt and J. W. O'Neil
ISBN: 978-1-62257-646-3

Chapter 97

Genetic Diversity of the Population of Russia: Gene Pool and Genegeography

Sergei Rychkov, Oksana Naumova, Alexei Evsyukov, Irina Morozova, Yuri Shneider and Olga Zhukova
Lab of Human Genetics, Vavilob Institute of General Genetics, Rusian Acad. of Science, Russia

Research Summary

Genetic differentiation of the population of Russia is investigated. The work is based on data about immuno-biochemical and molecular markers polymorphism in about 1,500 populations from 62 ethnoses belonging to six main linguistic families and having different cultural traditions. Genetic diversity is studied by cartographic and statistical methods and is presented in a form of genegeographical maps. The position of the Russian gene pool on the Eurasian background is described. The genetic relief of Russia is investigated, and main structure components are revealed in the gene pool. Analysis of these components from the ethno-historical point of view revealed their connection with different Eurasia regions (West and Central Europe, Central and East Asia).

In: Genetics Research Summaries
Editors: L. E. Hewitt and J. W. O'Neil

ISBN: 978-1-62257-646-3

Chapter 98

Genetic Variability within *Cypella fucata* Ravenna in Southern Brazil

***Évilin Giordana de Marco*[1], *Luana Olinda Tacuatiá*[1], *Lilian Eggers*[2], *Eliane Kaltchuk-Santos*[1] *and Tatiana Teixeira de Souza-Chies*[2]**

[1]Departament of Genetics
[2]Department of Botany, Universidade Federal do Rio Grande do Sul, Porto Alegre, Rio Grande do Sul, Brazil

Research Summary

Iridaceae is a relatively large family of monocots comprising over 2,030 species in 65-75 genera. *Cypella fucata* Ravenna is characterized as a perennial herb which presents bulbous and beautiful orange flowers that have ornamental value. The distribution of the species comprises Brazil, in the states of Rio Grande do Sul and Santa Catarina, and Uruguay. This study aims to compare two geographically distinct survey areas of *C. fucata* using molecular approaches and to offer a contribution to the knowledge of genetic variation of the species. *Cypella fucata* specimens were collected in the State of Rio Grande do Sul, Brazil, in two sites: the municipalities of Piratini (26 specimens) and Capão do Leão (28 specimens). Survey sites were localized along a road, and were 22 km distant from each other. Specimens were analyzed by ISSR-PCR (Inter Simple Sequence Repeats) since ISSR markers have a high capacity to reveal polymorphism and offer a great potential to determine intra- and interspecific levels of variation. Nine primers were tested, generating 201 fragments (bands) with sizes ranging from 150 bp to 2,000 bp and an average of 22 bands per primer. A matrix of presence and absence of fragments was constructed and the Jaccard's coefficient was calculated. A dendrogram based on these values was generated to reveal the genetic structure of both populations. The patterns were highly polymorphic within each collection site, with samples aggregated into two major groups, corresponding to the surveyed populations. In addition, ϕ_{ST} was calculated and may indicate some interpopulation gene flow (ϕ_{ST} =

0.0851) and an intermediate structure. The Nei's genetic distance showed a high identity between the two collection sites analyzed (98%). Since the sampled areas were near each other, our data may suggest that they in fact correspond to two subpopulations derived from a single original one. These data may indicate that *C. fucata* presents cross-pollination and the vegetative propagation does not play an important role in the maintenance of the populations. Specimens from other sites will be analyzed to confirm the mating system. This study is the first contribution to the knowledge of evolutionary aspects of this species.

In: Genetics Research Summaries
Editors: L. E. Hewitt and J. W. O'Neil
ISBN: 978-1-62257-646-3

Chapter 99

Genetic and Functional Diversity of Phosphate Solubilizing Fluorescent Pseudomonads and Their Simultaneous Role in Promotion of Plant Growth and Soil Health

K. Badri Narayanan, M. Jaharamma, G. Raman and N. Sakthivel*
Department of Biotechnology, School of Life Sciences, Pondicherry University,
Kalapet, Puducherry, India

Research Summary

Soil microbes that solubilize the insoluble phosphates play a vital role in maintaining soil fertility, plant health and subsequent enhancement of crop yield. Fluorescent pseudomonad group of bacteria are often predominant among bacterial species associated with the plant rhizosphere. Due to their innate capability for plant growth promotion, plant disease suppression and their potential for biodegradation of agricultural chemical pollutants, fluorescent pseudomonads have been a major focus for investigators around the world. In recent years, rich knowledge has been generated on diversity, functional potential of fluorescent pseudomonads. This chapter describes the genetic and functional diversity of fluorescent pseudomonads and their role in phosphate solubilization, biological control and soil fertility.

* Corresponding author: Dr. N. Sakthivel, Professor and Head,. E-mail address: puns2005@gmail.com. Tel: +91-413-2654430; fax: +91-413-2655255

In: Genetics Research Summaries
Editors: L. E. Hewitt and J. W. O'Neil

ISBN: 978-1-62257-646-3

Chapter 100

Genetic Diversity and Population Structure of Alpine Plants Endemic to Qinghai-Tibetan Plateau, with Implications for Conservation under Global Warming

Yupeng Geng[1], John Cram[2] and Yang Zhong[3]
[1]School of Life Sciences, Fudan University, Shanghai, China
[2]China-UK HUST-Rres Genetic Engineering and Genomics Joint Laboratory, Huazhong University of Science and Technology, Wuhan, China
[3]School of Life Sciences, Fudan University, Shanghai, and Institute of Biodiversity Science and Geobiology, Tibet University, Lhasa, China

Research Summary

The Qinghai-Tibetan Plateau is one of the most important centers of biodiversity for alpine species in the world and is among the areas that are most sensitive to global warming. Knowledge about population genetics is essential for understanding the dispersal ability and evolutionary potential of alpine species in a warming world. In this chapter, we review the genetic diversity and population structure of 19 alpine plant species endemic to the Qinghai-Tibetan Plateau. Generally, the population genetic variation can varygreatly among different species and the endangered species have much lower levels of genetic diversity than the co-occurring common species. Although a few species showed increased levels of genetic diversity along altitude, we dectected no significiant correlation between diversity and altitude in most species. In addition, the isolation-by-distance model cannot explain the spatial genetic structure in most alpine species that have been investigated, which may partially due to the discontinous distribution of alpine species shaped by complex geomorphology in Qinghai-Tibetan Plateau. The implications of these results for the conservation of alpine plants during global warming are discussed.

In: Genetics Research Summaries
Editors: L. E. Hewitt and J. W. O'Neil

ISBN: 978-1-62257-646-3

Chapter 101

Bayesian Inference under Complex Evolutionary Scenarios Using Microsatellite Markers: Multiple Divergence and Genetic Admixture Events in the Honey Bee, *Apis Mellifera*

Jean-Marie Cornuet*[*1], *Laurent Excoffier*[2], *Pierre Franck*[3] and *Arnaud Estoup*[1]

[1]Centre de Biologie et de Gestion des Populations, INRA, Campus International de Baillarguet, Saint-Gély-du-Fesc Cedex, France
[2]Computational and Molecular Population Genetics Lab (CMPG), Zoological Institute, University of Bern, Bern, Suisse
[3]UR1115 Plantes et Systèmes de culture Horticoles, INRA, France

Research Summary

Making inference from molecular data on the demographic parameters of complex evolutionary scenarios remains methodologically challenging. The approximate Bayesian computation (ABC) method has the potential to treat such scenarios (Beaumont *et al*.., 2002). We have developed a user-friendly methodological framework based on ABC that allows one to make inferences from microsatellite data under evolutionary scenarios including any combination of admixture, divergence and (discontinuous) effective population size variation events, and this for any number of populations. We illustrate here the potential of this methodological framework by making inferences on a complex scenario involving four *A.*

[*] Corresponding author: Jean-Marie Cornuet. Tel: +44 20 7594 3420; email: cornuet@supagro.inra.fr

mellifera populations sharing two divergence and two admixture events. Four groups of honey bee populations belonging to two genetic lineages (M and C) and genotyped at eight microsatellite loci have been analysed twice to evaluate estimation stability. In addition, mean relative bias and errors have been computed from 500 data sets simulated with known values of parameters (close to estimates on real data), showing that the order of magnitude of all parameters is correctly estimated. Time estimates of divergences between populations are compatible with previous estimates: -0.6 My for lineages M and C divergence and -0.2 My for French and Italian M lineage divergence. The estimated proportion of lineage M alleles in the subspecies *ligustica*, amounting to 12%, is intermediate between estimates obtained by two different methods. Furthermore, our ABC analysis allows decomposing the previous estimate of 35% of lineage M alleles in the recently admixed population as 23% from the local *mellifera* subspecies and 77%×12% (9.2%) from the imported *ligustica*, making a total of 32.2%. The most unexpected result concerns the time of the admixture of lineages M and C that gave rise to the subspecies *ligustica*. It is estimated at –2,000 years with an approximate credibility interval of (-1,000, -7,000).

In: Genetics Research Summaries
Editors: L. E. Hewitt and J. W. O'Neil

ISBN: 978-1-62257-646-3

Chapter 102

Geographic Structure of Craniometric Variation and the Estimates of Possible Dispersal Routes of Major Human Populations

Tsunehiko Hanihara *
Department of Anatomy and Biological Anthropology,
Saga Medical School, Saga, Japan

Research Summary

In the last decade, a near consensus has emerged in supporting single African origin of modern humans. However, the timing of dispersal out of Africa and the routes taken are far from obvious and focus of debate. In the present study, possible dispersal routes taken across Eurasia and finally New World and the Pacific were investigated using craniometric dataset consisting of 34 measurements. The degree of intra-regional variation shows that sub-Saharan Africans are the most diverse and that the diversity of non-Africans is negatively correlated with geographic distance to East Africa. The relationship between regional variation and geographic distance from sub-Saharan Africa tested by linear regression analysis supports a possible dispersal route proposed from the research of mtDNA haplotype variation, the Horn of Africa (the route across the Bab el Mandeb Strait) as a passageway in major human migration out of Africa. The results obtained support, moreover, the multiple migration hypothesis for the peopling of East/Northeast Asian region; mainly from central/western Asia with minor contribution from Southeast Asia. Nonlinear regression (exponential approximation) analysis using geographic distance measured along a hypothetical dispersal route shows that phenotypic similarity between populations decreases as the geographic

* Correspondence to: Tsunehiko Hanihara, Department of Anatomy and Biological Anthropology, Saga Medical School, 5-1-1 Nabeshima, Saga, 849-8501, Japan; E-mail: hanihara@cc.saga-u.ac.jp

distance increases. Such findings suggest that geographic distance is a primary and significant determinant of not only genetic but also craniometric variation between major human population groups. The present study illustrates that modern human cranial diversity patterns fit an evolutionary model of neutral expectation and a dispersal model of iterative founder effects with an African origin.

In: Genetics Research Summaries
Editors: L. E. Hewitt and J. W. O'Neil
ISBN: 978-1-62257-646-3

Chapter 103

Intra-Specific Genetic Variation in Mosses: A Novel Approach to Detect Environmental Changes

Valeria Spagnuolo, Stefano Terracciano and Simonetta Giordano
Dipartimento di Biologia Strutturale e Funzionale,
Università degli Studi di Napoli Federico II, Italy

Research Summary

Intra-specific genetic variation is considered an important factor for evaluating biodiversity; indeed, the higher genetic variation within a species, the higher its surviving ability. The loss of suitable habitats for moss species involves demographic decreases and genetic impoverishment. Mosses, have a short generation time compared to phanerogamic vegetation, particularly trees, and therefore may exhibit all these effects earlier, predicting the destiny of higher plant communities and the ongoing changes in natural landscapes. Indeed, intra-specific genetic variation in moss species may represent an ideal model system for investigating species fitness consequent to natural and man driven environmental changes, both at a local level, and at a large scale. At a local level these studies provide useful information for territory management since they promptly signal local environmental changes; whereas, over a large scale they highlight historical processes which have affected taxon origin, distribution, radiation, in relation to the main geological events. Genetic variation and structure within moss species is influenced by reproductive strategy and dispersal, giving information about gene exchange, occurrence of sexual reproduction, selfing/outcrossing rates. Demographic constraints and especially ongoing demographic fluctuations also concur to shape population genetic diversity and structure, evidencing phenomena such the relative importance of the founder effect, the occurrence of bottleneck and genetic drift. Moss genetic variation may highlight environmental disturbance caused both by natural events and by land use and human pressure. Among disturbances, habitat fragmentation is one of the most studied due to the increasing loss of suitable habitats for moss species. In general, it can be stated that intraspecific genetic variation in mosses reflect environmental gradients, with high amount of variation in natural environment, versus low level of variation in threatened environments.

The rapid transformation of the environment into a network of patches due to habitat fragmentation, and the increasing environmental disturbance, lead to a genetic erosion in isolated populations, with consequent increase of extinction risk. Thus, intraspecific genetic variation in mosses appears a suitable tracer of environmental disturbance due to the global ubiquity and the fast generation time of these plants.

In: Genetics Research Summaries
Editors: L. E. Hewitt and J. W. O'Neil
ISBN: 978-1-62257-646-3

Chapter 104

Introducing New Definition of Elderly in Japan

Yasuharu Tokuda *
Department of Medicine, St Luke's International Hospital
Chuo City, Tokyo, Japan

Research Summary

Japan is currently developing into an extraordinarily rapidly-aging society. The current official definition of elderly in Japan is "age of 65 years or older" and this aged population has exceeded beyond 20% in 2006. In addition, because of the lowest birth rate ever in Japan, the proportion of the younger aged group is also shrinking. These population dynamics in Japan can cause soaring healthcare costs, shrinking work forces, and possible collapsing pension programs. Since many of the Japanese elderly, around the age of 65-75 years, are still able to actively work and contribute to society and the average remaining life expectancy of Japanese at age 65 is about 23 years for females and 18 years for males, we have recently proposed re-defining elderly to "aged 75 or older". Older workers can remain productive and stay healthy longer. Work participation among seniors is beneficial for maintaining autonomy and quality of later life. The revised definition of elderly could make a huge and important impact in terms of maintaining the social and economic integrity in this country and many senior workers would remain active, stay healthier and are integrated in a rewarding social network.

* Phone and FAX 81-(0)3-5550-2426; Email: tokuyasu@orange.ocn.ne.jp

In: Genetics Research Summaries
Editors: L. E. Hewitt and J. W. O'Neil

ISBN: 978-1-62257-646-3

Chapter 105

Aging Puzzle

Mladen Davidovic, Dragoslav P. Milosevic, Nebojsa Despotovic, Jelena Vukcevic and Predrag Erceg
Clinical Department for Geriatrics, "Zvezdara" University Hospital
Belgrade, Serbia

Research Summary

In the past couple of centuries, scientists proposed great number of aging theories, but neither of them was completely satisfactory.

In the statistical sense we are faced with an even greater dilemma because of the presence of a large array of factors that affect ageing. Although most of these factors are well known to us at this point, just the fact of their innumerability complicates our approach to the issue at hand. These factors in the positive sense involve: starvation, Mediterranean diet, the life styles, activity and so on. In the negative sense we are talking about unhealthy habits, including the omission of the positive life styles listed above.

If we tried to sum up everything we know today about ageing into two sentences, as much as such simplifications can be dangerous, we would probably come up with the following two sentences or basic observations:

- First, caloric restriction diet does prolong life and secondly,
- Delay of the reproductive processes also prolongs life.
- Most recently it appears that high-predation risk in one group the guppies also prolong reproduction life span as life span in general.
- More and more authors are trying to explain the unknowns in the understanding of these observations about ageing, by adding the statement that there are two subgroups in the general population. Such statement is in no contradiction with the basic observations listed here. This acknowledgement of two subpopulations explains why there are numerous cases that can not be explained or defined or fitted in these basic observations about caloric restrictions and the delay of reproduction.

The two known options for the prolongation of life, that are listed here as basic observations about ageing, are applicable to the majority of the general population. The rest of the population, as much as it might seem awkward, can be considered as privileged. The primary groups, i.e. the majority of the population, are those that have evolved successfully, in other words there would be no evolution, if there wasn't for this group. The privileged group, "the selfish ones", are those that are genetically more stable. If the privileged group was the majority, human race would not have gone down the same evolutionary road.

In: Genetics Research Summaries
Editors: L. E. Hewitt and J. W. O'Neil

ISBN: 978-1-62257-646-3

Chapter 106

Toward the Mitigation of Glycation: A Practical Protocol

Nūn Sava-Siva Amen-Ra
Morgan State University, School of Community Health & Policy,
Baltimore, MD, US

Research Summary

The purpose of this chapter is to establish the theoretical and practical efficacy of a dietary/lifestyle regimen in reducing glycative damage and, by so doing, improving health and longevity. Known as the Amen Optimal Health Protocol, the regimen seeks to synergistically combine established experimental interventions with efficacy in lengthening lifespan or modulating physiological processes known to be operative in aging. Chief among such processes is glycation. Glycation is a chemical process wherein the carbonyl moieties of reducing sugars such as glucose or its metabolites react covalently with nitrogenous constituents of amino acids, proteins, or certain lipids. Through a series of complex chemical conversions various sugar-protein and sugar-lipid adducts eventuate in so-called advanced glycation endproducts (AGEs). It is these multifarious endproducts of glycation that have been determined to be deleterious to health and longevity.

In: Genetics Research Summaries
Editors: L. E. Hewitt and J. W. O'Neil

ISBN: 978-1-62257-646-3

Chapter 107

Prospects of a Longer Life Span Beyond the Beneficial Effects of a Healthy Lifestyle

*Giacinto Libertini**
Naples, Italy

Research Summary

Life span is limited by the effects of diseases and ageing. For the aim of a longer life span, it is indispensable that a rational analysis of the primary causes of these phenomena is not limited to the description of their physio-pathological mechanisms. If Dobzhansky's statement that "nothing in biology makes sense, except in the light of evolution" is true, it appears logical to maintain that evolution theory must be the main tool for such analysis.

From an evolutionary point of view, diseases are the predictable consequence of: 1) defects in the maintenance and transmission of genetic information; 2) alterations of the ecological niche to which the species is adapted (in particular, for our species, due to civilisation); 3) interactions with other species (bacteria, viruses, fungi, protozoa, parasitic worms, etc.); and 4) conditions for which the species is not adapted.

Moreover, evolution theory allows the paradoxical prediction that, in particular ecological conditions, kin selection favours a progressive fitness decline, usually, in its more evident manifestations, referred to as ageing and that must not be classified as a disease. This decline is genetically determined and regulated and is obtained with a progressive limitation of cell turnover through a sophisticated modulation of the telomere-telomerase system.

A lifestyle compatible with the ecological niche to which our species is adapted and good medical treatment permit the attainment of the highest life span defined by the genetic program of our species (within the limits of individual genetic peculiarities) but do not allow the overcoming of the maximal values of longevity defined by the same program.

* E-mail: giacinto.libertini@tin.it

To increase these values, it is indispensable to modify the genetic planning of ageing, with a different modulation of the telomere-telomerase system.

In principle, granted that this will be considered ethically acceptable, it is possible to propose a modification of that part of the genetic program that modulates ageing so that an unlimited survival is obtained, similar to the so-called "negligible senescence" observed for many animal and plant species.

A possible schedule to achieve this aim and the effects on human civilisation are outlined.

In: Genetics Research Summaries
Editors: L. E. Hewitt and J. W. O'Neil

ISBN: 978-1-62257-646-3

Chapter 108

Stress Resistance and Aging in Long-Lived Rodent Models

James M. Harper[1*] and Adam B. Salmon[2]
[1]Department of Pathology and Geriatrics Center,
University of Michigan School of Medicine, Ann Arbor, MI, US
[2]The Sam and Ann Barshop Institute for Longevity and Aging Studies,
University of Texas Health Science Center at San Antonio, San Antonio, TX, US

Research Summary

The aging process affects nearly every biochemical and physical system measured to date. Nevertheless, the mechanism(s) responsible for modulating the rate of aging have yet to be identified. A number of mechanistic theories have predicted that the accumulation of damage to cellular macromolecules could be key determinants of the rate of aging and the progression of age-related disease. These theories also predict that aging may be slowed by the attenuation of damage accumulation, a process that may be accomplished by enhanced resistance to stress.

Treatments that extend lifespan, for example the restriction of caloric intake, are known to reduce the accumulation of cellular damage, including oxidation and DNA mutation, as well as increase the resistance to numerous stressors such as oxidative stress, DNA damage and heat. In addition, genetic mutations that extend murine lifespan have consistently shown a corresponding increase in resistance to multiple types of stress. For instance, dwarf mutations, like the Snell and Ames dwarf, have been shown to extend lifespan by 40-60% while delaying multiple age-related pathologies. Dwarf mice also show an increased resistance to oxidative stress *in vivo*, as well as cellular resistance to multiple cytotoxic and metabolic insults.

In addition to these well-characterized mutant mouse models, wild-derived and free-living rodent populations are a virtually untapped resource in the context of stress resistance,

* Address correspondence to: James M. Harper, Ph.D. 109 Zina Pitcher Pl, Room 3005 BSRB Ann Arbor, MI 48109-2200 Email: jmharper@umich.edu Phone: 734-615-0494 FAX: 734-936-9220

aging and disease. Recent studies have shown that wild-derived mouse stocks are very different physiologically from laboratory mouse stocks and that they exhibit significant differences in life span. Work in these underutilized strains and other long-lived rodent models, has suggested that they also exhibit enhanced stress resistance, and especially an enhanced resistance to oxidative stress. Hence, it may be that among models of extended longevity, resistance to stress plays an important role in the maintenance of cellular homeostasis, thereby preserving cell and tissue function and delaying the progression of age-related phenotypes.

In: Genetics Research Summaries
Editors: L. E. Hewitt and J. W. O'Neil

ISBN: 978-1-62257-646-3

Chapter 109

Longevity-Associated Mitochondrial DNA 5178 C/A Polymorphism Modifies the Effect of Coffee Consumption on Glucose Tolerance in Middle-Aged Japanese Men

Akatsuki Kokaze[1,2,*], Mamoru Ishikawa[3,4], Naomi Matsunaga[2], Kanae Karita[2], Masao Yoshida[2], Tadahiro Ohtsu[1], Takako Shirasawa[1], Yahiro Haseba[1], Masao Satoh[4], Koji Teruya[5], Hiromi Hoshino[1] and Yutaka Takashima[2]

[1]Department of Public Health, Showa University School of Medicine, Shinagawa-ku, Tokyo, Japan
[2]Department of Public Health, Kyorin University School of Medicine, Mitaka-shi, Tokyo, Japan
[3]Mito Red Cross Hospital, Mito-shi, Ibaraki, Japan
[4]School of Medical Technology and Health, Faculty of Health and Medical Care, Saitama Medical University, Hidaka-shi, Saitama, Japan
[5]Department of Public Health, Kyorin University School of Health Sciences, Hachioji-shi, Tokyo, Japan

Research Summary

Mitochondrial DNA 5178 C/A (Mt5178 C/A), NADH dehydrogenase subunit-2 237 leucine/methionine (ND2-237 Leu/Met), polymorphism is reported to be associated with

* Correspondence to: Akatsuki Kokaze, Department of Public Health, Showa University School of Medicine, 1-5-8 Hatanodai, Shinagawa-ku, Tokyo 142-8555, Japan Tel: +81(3)3784-8133; Fax: +81(3)3784-7733; E-mail: akokaze@med.showa-u.ac.jp

longevity, blood pressure, serum lipid levels, fasting plasma glucose levels, glucose tolerance, serum uric acid levels, pulmonary function, hematological parameters, intraocular pressure, serum electrolyte levels, and serum protein fraction levels in the Japanese population. Mt5178A (ND2-237lMet) genotype reportedly provides resistance to diabetes and atherosclerotic diseases, such as myocardial infarction and cerebrovascular diseases. Several epidemiological studies have shown that habitual coffee consumption exerts a protective effect against abnormal glucose tolerance in a Japanese population. This study investigated whether longevity-associated Mt5178 C/A (ND2-237 Leu/Met) polymorphism modifies the effects of coffee consumption on the risk of abnormal glucose tolerance in middle-aged Japanese men. A total of 332 male subjects (age, 52.8 ± 7.8 years; mean ± SD) were selected from individuals visiting the hospital for regular medical check-ups. After adjustment, the odds ratio (OR) for abnormal glucose tolerance was significantly lower in subjects with Mt5178C (ND2-237Leu) who consume more than four cups of coffee daily compared with those who consume less than one cup of coffee daily (OR = 0.245; 95% CI: 0.066 - 0.914). On the other hand, the association between Mt5178A genotype and risk for abnormal glucose tolerance did not depend on coffee consumption. These results suggest that longevity-associated Mt5178 C/A (ND2-237 Leu/Met) polymorphism modifies the effects of coffee consumption on reduced abnormal glucose tolerance risk in middle-aged Japanese men. This novel diet-gene interaction may contribute to the individualized prevention of lifestyle-related adult-onset diseases and allow for healthy aging and longevity.

In: Genetics Research Summaries
Editors: L. E. Hewitt and J. W. O'Neil
ISBN: 978-1-62257-646-3

Chapter 110

Pesticides and Longevity

Sáenz de Cabezón Irigaray, Francisco Javier, Carvajal Montoya, Luz Dary and Moreno Grijalba, Fernando
Unidad de Protección de Cultivos, Departamento de Agricultura y Alimentación
Universidad de La Rioja, Spain

Research Summary

Most common and effective way of a pesticide to take effect is through direct ingestion via food, water or air. Once the active ingredient is inside the body of the target (pest) or non-target organism (humans, environment, wild and domesticated fauna), it can either affect vital processes killing the organism, or display single or multiple side effects that can have serious effects on other life processes affecting the longevity of the organism. After more than 7 decades depending on heavy use of chemical in agriculture, the rise of concern about how these active ingredients affect the physiology of different organisms, have developed ways to analyze and to describe the deleterious effects of pesticides on human and environmental health. In this chapter, we will discuss the interaction of pesticides, with target and non target physiological systems, giving emphasis on secondary effects (hormonal, developmental, genetic, general homeostatic changes) that can impair not only the life (longevity) of the affected organism but can have effects on successive generations. In this way, we will obtain an idea of how pesticide exposure can have various effects on human health, those effects can interact, synergize and lead to chronic damage at various physiological systems and organizational levels, as a result, if damages are important enough to impair or interfere with metabolic, homeostatic or/and functional responses, the target can view its lifespan, including reproductive outcomes, reduced.

In: Genetics Research Summaries
Editors: L. E. Hewitt and J. W. O'Neil
ISBN: 978-1-62257-646-3

Chapter 111

The Optimal Human Diet: Theoretical Foundations of the Amen Protocol

Nūn Sava-Siva Amen-Ra*
Amenta Press, Inc., Damascas, MD, US

Research Summary

The purpose of this treatise is to explicate the central elements of a theoretically optimal human diet. Such a diet is herein defined as one that is plausibly capable of increasing human longevity, decreasing disease susceptibility, augmenting mental acuity and generally facilitating the attainment and maintenance of an aesthetic ideal encompassing muscularity, leanness and juvenescence. The main parameters pertinent to the conceptualization of this theoretical diet are temporal, quantitative and qualitative. That is, consideration is given to the frequency with which food is to be ingested and the time at which it should be ingested, the amount of calories that should be consumed and the composition of the food so consumed. This treatise will review and analyze scientific studies and fundamental physiological principles that form the theoretical and empirical bases of the diet. Caloric restriction and cyclic fasting—experimental interventions known to increase maximum mammalian lifespan—are the empirical cornerstones of the theoretically formulated regimen. As to the qualitative component of the diet, consideration is given to foods whose chemical constituents are known to influence the same physiological processes modulated by caloric restriction and cyclic fasting. Such processes include facilitation of detoxification, augmentation of antioxidant defense, attenuation of inflammation and oxidative damage, inhibition of glycation and maintenance of membrane, protein and genomic integrity. Thus, a major tenet of my theoretical complex is what I herein term *selective synergism*. In the context of this treatise, selective synergism entails ascertainment of the mechanisms involved in lifespan

---

* AmentaPr@AmentaPress.com

prolongation and disease diminution secondary to caloric restriction and attempting to augment these effects dietetically. Finally, consideration shall be given to psychological factors influencing compliance with such a restrictive protocol, including experiential insights gleaned from the author's own self-imposed adoption of the protocol in question spanning a period of four years.

In: Genetics Research Summaries
Editors: L. E. Hewitt and J. W. O'Neil
ISBN: 978-1-62257-646-3

Chapter 112

Protective Effect of Calorie Restriction on Age-Induced Fibrosis

E. Chiarpotto[1], L. Castello[1], E. Bergamini[2] and G. Poli[1]
[1]Department of Clinical and Biological Sciences, University of Torino, Orbassano (TO), Italy
[2]Research Center on the Biology and Pathology of Aging, Department of Experimental Pathology, University of Pisa, Pisa, Italy

Research Summary

Starting from research by McCay et al., several studies have demonstrated that controlled calorie restriction (CR) exerts an anti-ageing effect in different organisms, from invertebrates to mammals. Observational studies suggest that CR also has beneficial effects on human longevity. However, the anti-aging mechanisms of CR are still not clearly understood. One mechanism might be protection against the age-associated increase of oxidative stress and consequent cellular damage.

In parallel, a role of oxidative stress in fibrosis induction and progression has been demonstrated in many human diseases, such as pneumoconiosis, interstitial pulmonary fibrosis, cystic fibrosis, cirrhosis, neurodegenerative diseases and atherosclerosis.

In fibrosis, fibroblasts or fibroblast-like cells are activated by various cytokines, among which transforming growth factor β1 (TGFβ1) is prominent, to proliferate and produce high levels of extracellular matrix and collagen. High TGFβ1 levels have been found in human diseases of different organs all characterized by excessive ECM deposition and marked fibrosis (cirrhosis, chronic hepatitis, glomerulonephritis, diabetic nephropathy, atherosclerosis, sclerodermia, pulmonary fibrosis). Fibrosis is also a constant distinctive feature in tissue aging.

TGFβ1 exerts its multiple biological activities through interaction with type I and type II receptors. Signaling to the nucleus is principally through cytoplasmic proteins of the Smad family, but in various cell types TGFβ1 also activates mitogen activated protein kinase

(MAPK) pathways, i.e. extracellular regulated kinase 1 and 2 (ERK1/2), c-Jun N-terminal kinase (JNK) and p38, leading to collagen type I synthesis through activation of the transcription factor activator protein 1 (AP-1).

In this context, it is of interest to study the possible protective effect of CR against the onset of fibrosis in the frame of tissue and soma aging.

In this connection, our group has shown that, with aging, there is an increase of oxidative stress in rat aorta, with a progressive hoarding of biologically-active end-products, in particular 4-hydroxy-2,3-nonenal (HNE). Moreover, the increase of oxidative stress with aging is accompanied by increased fibrosis in terms of TGFβ1 and collagen levels. CR protects against both phenomena.

With regard to possible protective mechanisms of CR against fibrosclerosis, we believe they may be closely connected to the reduction of oxidative stress: by decreasing HNE levels in older rat aortae, CR significantly decreases JNK and p38 activity. Since JNK is central for AP-1 activation, by negatively modulating JNK, CR also prevents AP-1 activation and consequently down-regulates transcription of AP-1-dependent genes, including TGFβ1, vimentin and collagen.

The possibility of controlling the fibrogenesis process by modifying dietary habits opens new nutritional horizons in the prevention and treatment of several pathological processes characterized by excessive fibrosis. However, since it seems difficult to transpose animal CR model as such to man, interest in natural and/or pharmacological CR mimetic molecules is increasing.

In: Genetics Research Summaries
Editors: L. E. Hewitt and J. W. O'Neil
ISBN: 978-1-62257-646-3

Chapter 113

Engineered Natural Longevity-Enhancing Interventions

Arkadi F. Prokopov[*1] *and Tamara N. Voronina*[2]
[1]Integrative Medicine, Mallorca, Spain
[2]Ukraine
Chiltern Street, London, UK

Research Summary

Currently developing engineered biogerontological interventions promise to radically extend healthy life span. Nevertheless, the exploration of natural strategies that underlie longevity and resistance to age-related diseases in exceptionally long-living mammals, such as the bowhead whale (BW, *Balena mysticetus*) and naked mole rat (NMR, *Heterocephalus glaber*) is also rational. This paper is an attempt to analyze the advantages and limitations of these strategies, and to elucidate their existing derived applications, as well as to estimate their future prospects in the practice of biogerontology. Two natural longevity-modulating strategies (intermittent calorie/nutrient restriction—ICR, and intermittent oxygen restriction—IOR) utilize the universal development-and adaptational genetic programs, evolutionary "preinstalled" in all aerobic organisms. ICR and IOR synergistically diminish the basal level of mitochondria-dependent oxidative stress that is supposed to be the key factor, modulating life span and resistance to age-related diseases in aerobes.

ICR and IOR employ a common mitochondria-rejuvenating pathway, *mitoptosis*—a selective elimination of excessively ROS-producing mitochondria in the cells. Mitoptosis is a natural process that maintains the "quality" of mitochondria in the female germinal cells during early embryogenesis, and can be stimulated and maintained by IOR and ICR also in the postmitotic cells of adult organisms. Behaviorally induced continuous mitoptosis seems to be the key mechanism responsible for exceptional longevity and resistance to age-related diseases in BW and NMR.

[*] Email: ark1860@gmail.com

Additionally, ICR and IOR influence the longevity and tempo of development of age-related diseases via several mitochondria-independent pathways, such as suppressed protein glycation, enhanced DNA repair, accelerated protein turnover, stimulation of EPO, GH, HSP70 and other functional proteins. In addition, IOR distinctively intensifies stem-cell-dependent tissue repair. The unified positive effect of IOR and ICR on the cells and organisms manifests in enhanced genome stability and increased non-specific resistance to multiple stressors. Various forms of ICR and IOR have an impressively positive account of empiric and evidence-based use in the health-improving practices of various human cultures. Recently developed engineered techniques, such as short-termed, or alternate-day fasting (ADF, derivative of ICR) and intermittent hypoxic training/therapy (IHT, derivative of IOR) induce measurable mitochondrial and systemic rejuvenation.The vitamin-nutriceutical supplementation synergistically enhances the efficiency of IOR and ICR. Further development of engineered ICR and IOR protocols should facilitate their advanced clinical implementation and user-friendly, self-help applications.

In: Genetics Research Summaries
Editors: L. E. Hewitt and J. W. O'Neil
ISBN: 978-1-62257-646-3

Chapter 114

Be Young in Mind – Against Aging and Alzheimer's Disease

Kun Zou*
Department of Neuroscience, School of Pharmacy, Iwate Medical University, Nishitokuda, Yahaba, Iwate, Japan
Department of Alzheimer's Disease Research, National Institute for Longevity Sciences, NCGG, Morioka, Obu, Aichi, Japan

Research Summary

Life expectancy around the world has increased steadily for nearly 200 years. The continuing increase in life expectancy has made us face age-related diseases that we overlooked in the nineteenth and early twentieth centuries. Age is by far the biggest risk factor for a wide range of clinical conditions that are prevalent today. Although cardiovascular diseases and cancers still remain the leading causes of death for both men and women among age-related diseases, Alzheimer's disease has been in the spotlight recently because the number of its patients is increasing at an unexpected speed. Onset of Alzheimer's disease begins with short-term memory loss in its early stage and gradually the sufferer loses minor, and then major bodily functions, until *death* occurs. Unlike cardiovascular diseases and cancers, current knowledge of the processes leading to Alzheimer's disease is still limited, and very few effective treatments are available. Many treatment strategies have reached the clinical trial stage and some hopeful clinical trials failed to improve the survival or the cognitive ability of Alzheimer's patients. Up to now, clinically validated treatments for Alzheimer's disease remain confined to symptomatic interventions and recent studies suggest that frequent cognitive activity in old age has been associated with reduced risk of Alzheimer's disease. Thus, preventive strategies have become important in our daily life against aging and Alzheimer's disease.

* Tel: 81 19 698 1820; Fax: 81 19 698 1864; Email: kunzou@iwate-med.ac.jp

In: Genetics Research Summaries
Editors: L. E. Hewitt and J. W. O'Neil

ISBN: 978-1-62257-646-3

Chapter 115

Understanding Aging: Prejudices and Misconceptions

Carles Zafon
Division of Endocrinology, Hospital General Vall d'Hebron
Barcelona, Spain

Research Summary

Current evolutionary biology has rejected any theory that supports a selective purpose for aging. However, recent findings have forced a reconsideration of this concept and an effort to update old statements that have been made. In accordance, several evolutionary propositions have been suggested to justify the broad presence of aging in the animal kingdom and the wide range of longevity. Nevertheless, many of these alternative hypotheses have been dismissed using a small set of arguments that have been systematically repeated without, in some cases, any compelling proof of evidence. Moreover, there exists a variety of preconceptions and biases that could hinder the progression in theoretical thought about aging and longevity. The aim of this "author comment" is to expose some of these preconceptions to generate a productive debate which can enrich the progress in the field of evolutionary biology of aging. In particular, I will discuss three critical points that have not usually been considered in related papers. The first point is the "emotional bias" observed when authors argue the purpose of aging. The second point (When does aging begin?) pertains to the beginning of aging. The third topic (The old watchmaker) discusses the meaning of the aging process from an evolutionary perspective, and the evolutionary origin of the aging strategy.

In: Genetics Research Summaries
Editors: L. E. Hewitt and J. W. O'Neil
ISBN: 978-1-62257-646-3

Chapter 116

Longevity: Genetics, Diet and Disease

E. Naumova, I. Atanasova, M. Ivanova and G. Pawelec
Central Laboratory of Clinical Immunology,
University Hospital, Bulgaria

Research Summary

The aging process is very complex. Human longevity is a multifactorial trait which is determined by genetic and environmental factors and the interaction of "disease" processes with "intrinsic" ageing processes. Twin and family studies imply that up to 25% of human lifespan is heritable and increases in long-lived people to 36-40%. Longevity is a complex phenotype associated with different genes and gene polymorphisms, some of which are involved in metabolic pathways and endocrine system functioning, which affect diet-related health risks. For example, about 10 loci in the human genome have now been found that seem to confer susceptibility to type 1 diabetes mellitus, where insulin deficiency is due to autoimmune beta cell destruction. Nutrition is one of the hypothetical disease-triggering factors and low and special carbohydrate diet is obligatory for disease control. The best studied longevity pathway involves the genes regulating insulin and insulin-like growth factor-1 (IGF-1) signaling, as well as environmental effects on gene expression. Some of these genes are responsible also for oxidative stress, inflammation and tumor suppression. Another conventional group of genes currently under intense investigation are those involved in innate and adaptive immune responses. The pathogenesis of some of the most frequent human age-related diseases, like metabolic syndrome, diabetes mellitus type 2, and atherosclerosis is associated with a change in the circulating levels of C-reactive protein, tumor necrosis factor-α, interleukin-6, leptin, resistin, adiponectin, omentin, visfatin, and expression of CD40, - CD40 ligand, E-selectin, ICAM-1, and VCAM-1. Some of these genes may be responsible for mediating the beneficial effects of caloric restriction by regulating energy balance as well as stress resistance and oxidative damage. Additionally, other genes such as those encoding TLR, CCR5, KIR and MBL2 have also been associated with different age-related diseases and therefore with human longevity.

There are rare monogenic human diseases which are models for accelerated aging and progeria. Others are caused by inborn errors of metabolism: inherited traits that are due to a mutation in a metabolic enzyme, mutations in regulatory proteins and in transport mechanisms. Some metabolic disorders are due to structural chromosome aberrations and genomic imprinting, such as life-threatening obesity in Prader-Willi and Angelman syndrome.

In conclusion, this review will summarize current information on genome-wide dietary target genes, regulatory pathways affecting human homeostasis and diseases, and the role of immune response genes and particularly genes of innate immunity with an impact on lifespan.

In: Genetics Research Summaries
Editors: L. E. Hewitt and J. W. O'Neil
ISBN: 978-1-62257-646-3

Chapter 117

Effects of Chewing Ability on Longevity

Toshihiro Ansai[1*], Yutaka Takata[2] and Tadamichi Takehara[1]
[1]Division of Community Oral Health Science, Kyushu Dental College
[2]Division of General Internal Medicine, Kyushu Dental College, Japan

Research Summary

The oral cavity in humans has a variety of functions, such as eating, swallowing, and speaking, as well as chewing, which is one of its most important roles. Although the association of chronic periodontitis with systemic health has become an important issue, few studies have been conducted regarding the association between chewing ability and systemic health status. Recently, we conducted a population-based prospective study to determine the association between chewing ability and longevity in 697 (277 males, 420 females) community-dwelling elderly subjects in Japan, and herein present some of those results. Chewing ability was assessed based on the number of types of food each subject reported able to chew in answers to a questionnaire, which showed that individuals with higher chewing ability lived longer, after adjusting for various confounding factors (multivariate hazard ratio: 2.6; 95% confidence interval: 1.2-5.5). When the associations between causes of death and chewing ability were further analyzed, subjects reporting the lowest numbers of chewable foods were associated with a higher risk of mortality from cardiovascular disease (CVD) as compared with those able to chew all the types of food surveyed (multivariate hazard ratio: 5.1; 95% confidence interval: 1.1-23.8). On the other hand, there was no significant association between chewing ability and other causes of death including cancer. Our results indicate that maintenance of chewing ability in elderly individuals contributes to longevity. In addition, we discuss the significance of masticatory function in light of nutrition status.

* Correspondence to: Toshihiro Ansai, DDS, Ph.D., Division of Community Oral Health Science, Department of Health Promotion, Kyushu Dental College, 2-6-1 Manazuru, Kokurakita-ku, Kitakyushu 803-8580, Japan Tel: +81-93-582-1131 (ext. 2102). Fax: +81-93-591-7736. E-mail: ansai@kyu-dent.ac.jp

In: Genetics Research Summaries
Editors: L. E. Hewitt and J. W. O'Neil

ISBN: 978-1-62257-646-3

Chapter 118

Research on Aging and Longevity in the Parthenogenetic Marbled Crayfish, with Special Emphasis on Stochastic Developmental Variation, Allocation of Metabolic Resources, Regeneration, and Social Stress

*Günter Vogt**
Department of Zoology, University of Heidelberg,
Heidelberg, Germany

Research Summary

This article presents first results on aging and longevity in the marbled crayfish, an isogenic invertebrate with indeterminate growth. The marbled crayfish is the only known parthenogenetic species of more than 10.000 decapod crustaceans and has a maximal life span of roughly 3.5 years. Its main advantages, aside from genetically identical offspring and lifelong growth, are the alternation of growth and reproduction phases, a high regeneration capacity and easy handling in the laboratory. In a group of seven genetically identical batch-mates life span varied from 437 to 910 days although the sibs were communally reared and fed ad libitum with the same pellet food. In the same group there was no clear-cut relationship between longevity and growth or reproduction frequency. However, the specimen with the lowest life span showed fast growth, early onset of reproduction, and short time intervals between reproduction cycles. Damages like loss of appendages were repaired and did not negatively affect longevity. Social stress, in contrast, shortened life expectancy. The

* E-mail: vogt@zoo.uni-heidelberg.de

biological peculiarities of the marbled crayfish and the data obtained so far argue for a more intense use of this animal in research on aging and longevity.

In: Genetics Research Summaries
Editors: L. E. Hewitt and J. W. O'Neil
ISBN: 978-1-62257-646-3

Chapter 119

Genetic Background for Longevity in Livestock Species: A Review

Joaquim Casellas
Genètica i Millora Animal, IRTA-Lleida, Lleida, Spain

Research Summary

Longevity is a trait of major interest in livestock with important implications on both economic and animal welfare grounds. The relevance of longevity within the animal production framework was highlighted during the twentieth century with a plethora of studies focusing on environmental sources of variation. Although the specific molecular mechanisms involved in the longevity of domestic species remains unclear, recent studies have provided plenty of evidence of genetic determinism for aging-related traits in livestock. Indeed, longevity must be viewed as the final expression of a dynamic system governed by intrinsic (e.g. immunology, reproduction, and growth) and extrinsic determinants (e.g. temperature, food, and predators), in which every intrinsic determinant (and some extrinsic determinants, such as resistance to cold temperatures) is virtually modulated by genes. As the influence of genes is inextricably intertwined with longevity, it is virtually impossible to consider the broad determinants of longevity without a basic understanding of the underlying genetics.

Although longevity is not a conceptually new trait in livestock production, several advances relating to genetic background and genetic mechanisms have been achieved during the last few decades. The earliest investigations focused on non-additive genetic effects (i.e. inbreeding depression and crossbreeding/heterosis) that have a non-negligible impact on livestock longevity. Nevertheless, the recent development of specific analytical models has allowed a straightforward characterization of additive genetic sources of variation with moderate to low heritabilities in all species, and an increasing number of quantitative trait loci and candidate genes that influence the aging process. These models have resulted in systematic international genetic evaluations for cow longevity in dairy cattle and in some local breeding programs in other species and breeds. This has produced the first evidence of genetic progress due to selection programs. Future research into livestock longevity will focus on new technological platforms such as massive single nucleotide polymorphism genotyping and gene expression microarray in order to provide a more detailed characterization of the aging process in our domestic breeds. Within this context, the genetics of longevity in livestock species encompasses not only estimations of preliminary additive and non-additive genetic effects, but also national/international breeding programs and the latest technologies for addressing the genetic mechanisms that determine livestock lifespan.

In: Genetics Research Summaries
Editors: L. E. Hewitt and J. W. O'Neil

ISBN: 978-1-62257-646-3

Chapter 120

Genetic and Environmental Factors Affecting Inflammatory Responses to Infection

C. C. Blackwell, S. M. Moscovis, S. T. Hall, J. Stuart, C. J. Ashhurst-Smith, J. Roberts-Thomson, D. Yarnold and R. J. Scott
University of Newcastle, Newcastle, NSW, Australia
John Hunter Hospital, New Lambton, NSW, Australia

Research Summary

The incidence of notifiable infectious diseases is higher in Indigenous Australians compared with the non-Indigenous population, and there are higher incidences of heart and kidney diseases for which infection and inflammation are thought to play important roles in their pathogenesis. These discrepancies in health have been attributed mainly to poor socioeconomic conditions in which many Indigenous families live. New evidence is emerging that indicates there are significant genetic differences which might contribute to susceptibility to or severity of infections. The objective of this review is to examine these differences in relation to classical risk factors to determine potential interactions between genetic background and environmental conditions that contribute to susceptibility to these diseases.

For at least 40,000 years, epidemic infections considered to be major selective pressures for many genetic characteristics associated with immunity and infection were absent among Indigenous Australians. Until the late 18th century, because of its geographic isolation, the population was not regularly exposed to new infectious agents such as those which spread through Europe and Asia. In addition to its geographic isolation, Australia had other factors that reduced the likelihood of acute epidemic diseases. Indigenous Australians were nomadic and often relocated once food sources were depleted. This did not allow large amounts of human waste to accumulate and become a significant reservoir of infection. The low population density, scattered tribal groups, and the limited rate of travel reduced swift spread of any diseases with short incubation periods. If an individual or group became infected, the epidemic would most likely be restricted and not transmitted to another group. Except for the north of the country, there was little exposure to invasion or to "foreign trade routes" such as those between Europe and Asia along which epidemics traveled. The lack of domesticated animals in Indigenous tribes might also have been a contributing factor. Animals, like humans, require large populations to maintain epidemics. The development of agriculture and herding provided opportunities for transfer of pathogens from domesticated animals to humans. In addition, there is evidence that many of the most common epidemic viral and bacterial diseases are derived from those that affect domesticated animals: measles and tuberculosis from cattle; influenza from pigs and poultry; whooping cough from pigs and dogs.

Initial studies of DNA samples from a remote Indigenous Australian community indicate that their cytokine gene polymorphisms differ significantly from those of non-Indigenous, mainly European/Caucasian Australians. The arrival of Europeans introduced not only epidemic diseases characteristic of agrarian societies but also the conditions for their survival and spread. In addition to new microorganisms, Europeans also introduced changes in diet that included tobacco use and alcohol consumption.

In: Genetics Research Summaries
Editors: L. E. Hewitt and J. W. O'Neil

ISBN: 978-1-62257-646-3

Chapter 121

Mammalian Cell Sporulation – A Possible Link to Disease and Aging

***Yonnie Wu*[1], *Kyle Heim*[2] *and Erick Vasquez*[3]**
[1]Clemson University Genomics Institute
[2]Department of Biochemistry and Genetics,
[3]Department of Chemical and Biomolecular Engineering,
Clemson University, SC, US

Research Summary

This chapter reviews the physiochemical properties of mammalian spores and straw cells, which develop in vitro from the transformation of normal mammalian cells exposed to desiccative conditions. We hypothesize that this transformation is directly linked with aging and aging related diseases. Unlike many of the current damage related aging theories involving damage to DNA and protein molecules (leading to improper gene expression and protein function), this transformation is an active process at the cellular level. The metamorphosis of a normal cell into a mammalian spore is an inherited survival mechanism that can cause the loss of the body's most basic life-sustaining unit, the cell.

Mammalian spores are extremely prevalent. It is estimated that a typical person currently possesses billions of spores in their blood; the amount of spores may fluctuate in a day, perhaps influenced by physiological conditions such as diet and exercise that affect the overall sporulation stress. These mammalian spores form long extensive and extremely hydrophilic filamentous network that could contribute to blood viscosity. Blood cholesterol level may be increased from sporulation of billions of cells per day.

The metamorphosis of a mammalian spore from a normal mammalian cell is marked by a variety of distinct physical changes. First, the nuclei is broken into many pieces and reassembled within a tiny tube shaped structure. Second, tubular walls are constructed. Finally, upon the reintroduction of water, mammalian spores revert to normal mammalian cells and begin to divide. The early steps of this transformation resemble the development of apoptotic bodies. However, instead of destroying itself, the cell enters a robust spore-like cell survival phase, allowing it to remain viable under severe external conditions. Mammalian spores construct a strong cell wall using sulfated-glucose polymers and glycosylated acidic proteins, allowing it to withstand harsh treatment such as UV-C radiation.

A direct inverse relationship is displayed between the number of mammalian spores present in the blood of 9 domestic animals and these animals' average lifespan. The number of mammalian spores present in the heart and lung tissues, as well as excreted in the urine, also increases with the age of the mammal. This increased production of mammalian spores may play a role in disrupting normal cell and organ function. The large number of mammalian spores produced by a variety of tissues makes this transformation appear to be a natural physiological process embedded within a cell's environmental response machinery. The universal nature of mammalian spores and their mobility within the bloodstream allow them to affect many parts of the body. The mobile character of these cells displays a potential link to longevity and the development of aging-related diseases, including the metastasis of various cancers.

Water is the most essential component for cell survival. Throughout evolution organisms have developed many novel adaptations for particular environmental deficiencies and conditions. The formation of spores may be an early form of environmental adaptation for single celled organisms. However, in multicellular organisms the mammalian spore survival mechanism may not be beneficial because the organism loses necessary functional cells. Inhibiting mammalian spore biogenesis would prevent the depletion of normal healthy cells, as well as the spread of potentially harmful cells, and therefore improve mammalian longevity. Intervention of sporulation using membrane interacting molecules is discussed. Exogenous phosphotidylcholines (DOPC) significantly promote dehydration-induced sporulation while other phospholipids (DOPG, DOPE, DOPA, and DOPS) are ineffective.

In: Genetics Research Summaries
Editors: L. E. Hewitt and J. W. O'Neil
ISBN: 978-1-62257-646-3

Chapter 122

Carbon Nanotubes in Nucleic Acids and Affinity Biosensors

***Tibor Hianik*[1*] *and Gennady Evtugyn*[2]**

[1]Department of Nuclear Physics and Biophysics, Comenius University, Bratislava, Slovakia
[2]Analytical Chemistry Department, Kazan State University, Kazan, Russian Federation

Research Summary

Carbon nanotubes are of considerable interest in the development of biosensors for detection of DNA hybridization and damage, for affinity biosensors based on DNA/RNA aptamers or antibodies and for detection pathogens. Due to their unique properties, the carbon nanotubes can serve as an efficient support for immobilisation of nucleic acids and proteins. This chapter consists of an introduction into the structural and physical properties of carbon nanotubes and their applications in biosensing. The focus is on development biosensors for detection of DNA hybridization, single point mutations and polymorphisms as well as on affinity biosensors based on proteins and on DNA/RNA aptamers. Special focus on DNA aptamers is due to the unique properties of these new biopolymers that demonstrate substantial advantage over antibodies. While the sensitivity of aptamer-based biosensors is similar or even higher in comparison with that of antibodies, the aptamers can be regenerated without lost of affinity. Aptamers are rather flexible, which allow the applying of simple molecular engineering methods to construct aptamers with improved sensitivity. Immobilisation of aptamers on carbon nanotubes further improve the biosensor properties, which is reflected in better detection limits for various analytes in comparison with biosensors based on conventional immobilisation methods.

[*] Corresponding author: e-mail: tibor.hianik@fmph.uniba.sk, Tel: +421-2-60295683; Fax: +421-2-65426774;

In: Genetics Research Summaries
Editors: L. E. Hewitt and J. W. O'Neil

ISBN: 978-1-62257-646-3

Chapter 123

Bringing Tissue Identification into the 21st Century: mRNA Analysis as the Next Molecular Biology Revolution in Forensic Science?

Trisha L. Noreault-Conti and Eric Buel*
Vermont Forensic Laboratory,
Department of Public Safety, Waterbury, VT, US

Research Summary

DNA analysis using advanced molecular biological tools is universally used to unlock identifying information contained in crime scene stains. These advances have allowed scientists to provide critical information to solve crimes and allow the criminal justice community to convict the guilty and exonerate the innocent. Scientific research continues to develop new automated technologies and methods to yield more information from limited samples allowing scientists to optimize time and effort. Although the changes realized over the last 20 years have been remarkable, only small changes have occurred in the identification of the tissue source of a biological stain. Considerable research has been conducted to offer an alternative to these classical identification methods through the use of mRNA analysis. The application of mRNA into the forensic laboratory has not yet taken hold, but offers a new approach that could unequivocally identify difficult to characterize stains and allow the use of molecular biological tools for the identification of stains found at a crime scene. The nature of the research in this chapter is to identify mRNA transcripts that will definitively identify the tissue of origin, determine if such transcripts survive the typical environmental insults that forensic samples may encounter, and develop rapid assays using small amounts of sample. The stability of RNA was evaluated over time using real-time TaqMan®-based PCR assays.

* Corresponding author: E-mail: ebuel@dps.state.vt.us

Once RNA was shown to be stable in samples aged up to 4 years, 2-3 tissue-specific transcripts were identified for a variety of stains. In addition to specificity, the sensitivity of the assays was also determined using different sized stains. Once candidates were shown to be tissue-specific, the Plexor® One-Step qRT-PCR System was used to develop multiplex assays. This system allows the multiplexing of multiple RNAs in one assay, thus reducing the amount of sample needed and time of analysis. Through collaboration with Promega, a quick, one-tube seminal fluid-saliva screening assay was designed, since analysis of these fluids is a major task for all forensic laboratories. Additionally, a RNA/DNA co-isolation technique was developed which effectively extracts both nucleic acids, thus removing the requirement for separate extractions and allows DNA profiling from the exact stain in which RNA testing was performed. In conjunction with Promega, a likely outcome is a real-time PCR screening assay that determines with certainty whether a stain contains seminal fluid and/or saliva, an invaluable tool to the forensic community.

In: Genetics Research Summaries
Editors: L. E. Hewitt and J. W. O'Neil
ISBN: 978-1-62257-646-3

Chapter 124

The Potential Role of Active Methylation Cycle into Leukemic Cell Differentiation

Ioannis S. Vizirianakis**[*#]**, Sotirios S. Tezias and Asterios S. Tsiftsoglou
Laboratory of Pharmacology, Department of Pharmaceutical Sciences,
Aristotle University of Thessaloniki, Thessaloniki, Greece

Research Summary

Malignant cells exhibit metabolic alterations affecting growth, lack to respond to physiological stimuli and blockade of differentiation and apoptosis. Selective growth advantage on cells acquired by transformation influences biosynthetic enzymes and alters the internal pools of nucleotides. During the course of studying the induction of erythroid differentiation of murine erythroleukemia (MEL) cells, we observed that agents that inhibit the active methylation cycle, (N^6-methyladenosine, neplanocin A, 3-deazaneplanocin A and cycloleucine), blocked initiation of commitment, erased "memory" response to previous exposure to chemical inducers, altered the expression profile of specific genes and abrogated hemoglobin production in inducer-treated MEL cells presumably by interrupting a central process required for terminal erythroid maturation. In an effort to delineate the mechanism(s) by which such methylation inhibitors abrogate the developmental program of MEL cells, we have shown that exposure to these methylation inhibitors led to cytoplasmic accumulation of relatively small discrete RNA transcripts lacking poly(A) tail. These RNAs were hybridized

[*] Corresponding author: Laboratory of Pharmacology, Department of Pharmaceutical Sciences, Aristotle University of Thessaloniki, GR-54124 Thessaloniki, Greece; tel.: +30-2310-997658; Fax: +30-2310-997645; Email: <ivizir@pharm.auth.gr>.

[#] The authors' experimental data presented in this chapter represent their work published earlier in different journals cited at the reference section.

with ^{32}P-labeled probes derived from 3'-end flanking sequences of the mouse β^{major} globin gene. Structural and functional analysis of this DNA region have revealed a B1 repeat element as well as consensus sequences recognized by transcription factors GATA-1, AP1/NF-E2 and erythroid Krüppel-like factor (EKLF). Successful cloning, characterization and identification of genomic DNA sequence encoding these RNA species would shed some light on how do the active methylation cycle inhibitors activate their biosynthesis and subsequently whether these RNA molecules contribute to the blockade of differentiation of MEL cells by hypomethylating agents.

In: Genetics Research Summaries
Editors: L. E. Hewitt and J. W. O'Neil
ISBN: 978-1-62257-646-3

Chapter 125

Mitochondrial DNA Variation, Mitochondrial Haplogroups and Neurodegeneration

M. Mancuso, C. Carlesi, D. Orsucci and G. Siciliano
Department of Neuroscience, Neurological Clinic,
University of Pisa, Italy

Research Summary

Mitochondria, the powerhouse of the cell, play a critical role in multiple metabolic processes as well as in the apoptotic pathways. Many lines of evidence suggest that mitochondria and its DNA (mtDNA) have a central role in ageing-related neurodegenerative diseases. Moreover, there is a long history of investigations on mitochondria aimed at identifying genetic markers that would relate to aging or neurodegenerative diseases. In this chapter, we highlight some of the major neurodegenerative disorders and discuss the role of the mtDNA variations and haplogroups in the pathogenetic cascade leading to these diseases.

In: Genetics Research Summaries
Editors: L. E. Hewitt and J. W. O'Neil
ISBN: 978-1-62257-646-3

Chapter 126

The Continuing Evolution of Forensic DNA Databases

Simon J. Walsh*[*], *John S. Buckleton and Olivier Ribaux
Forensic & Data Centres, Australian Federal Police, Canberra, ACT, Australia
ESR Ltd., Auckland, New Zealand
Institut de Police Scientifique, Universite de Lausanne, Switzerland

Research Summary

In terms of forensic-specific applications of DNA technology, few have had a more profound effect on the field than the launching of databases around the world. The expansion of this phenomenon has not only required an appropriate level of scientific sophistication, but also unprecedented legislative, political and financial backing. In practical terms DNA database implementation and operation has brought massive increases in case submissions and a demonstrable change in case and evidence profiles. Forensic DNA databases also remain high profile and retain some element of controversy, typically in relation to socio-legal issues. Despite this impressive impact, it is important to acknowledge that forensic DNA database applications are still young, still only partially developed and understood and still requiring of our attention, monitoring and effort to ensure an ongoing positive contribution.

Fresh complexities are emerging in the management and development of DNA databases. Some of these problems have been examined in detail in this chapter, and, new solutions that allow cross-comparison of database performance are discussed. These methods should assist, not only in identifying present strengths and weaknesses but also in enhancing future database performance and effectiveness. Effective and creative database management has to be seen as a key priority for the forensic community. Achieving it will require a concerted effort of oversight that scientists themselves must be prepared to participate in as it is crucial to maximising the benefits of their work.

[*] Corresponding authors: E-mail: simonjoseph.walsh@afp.gov.au

Forensic science has traditionally focussed on the production of evidentiary outcomes – retrospectively, and with a view to confirming or refuting an already-held belief. DNA databases also offer, however, the potential for proactive forensic contributions to investigations. Delivering these requires functional and philosophical adjustments to the process by which forensic outcomes are produced. This chapter also outlines specific steps that need to be taken to progress beyond the traditional operational framework to one that is also conducive to effective policing and crime resolution.

In: Genetics Research Summaries
Editors: L. E. Hewitt and J. W. O'Neil

ISBN: 978-1-62257-646-3

Chapter 127

Transmitochondrial Mice in Studying MtDNA Distribution and Transmission

Mikhail G. Bass, Vassilina A. Sokolova, Maria E. Kustova, Oksana V. Kidgotko, Faina M. Zakharova and Vadim B. Vasilyev

Research Institute of Experimental Medicine,
Russian Academy of Medical Sciences, St. Petersburg, Russia

Research Summary

Obtaining reliable animal models of human OXPHOS (oxidative phosphorylation) diseases caused by accumulation of mutant mtDNA in cells of certain tissues might be a noticeable step towards understanding their pathogenesis. Despite some attempts, so far no successful approach to the targeted delivery of mutant mtDNA to certain tissues/organs has been elaborated in experiments on animals. We have elaborated a model that allows studying some features of foreign mtDNA distribution at different stages of embryonic development and in organs of transmitochondrial mice developed from zygotes injected with human mitochondria. All pre-implantation embryos and some neonate animals were carrying copies of human mitochondrial genome along with murine mtDNA, which is an indication of artificial heteroplasmy. This feature persisted in the progeny of transmitochondrial animals obtained from injected zygotes i.e. among the mice in generations F_1-F_3 at least 14% were human mtDNA-carriers inheriting foreign genome along the maternal lineage. Only some transmitochondrial embryos had human mtDNA in all the blastomeres. Likewise it was found only in some organs of transmitochondrial mice. Statistical elaboration and Monte Carlo simulation taking into account Markovian process showed, firstly, that human mtDNA was unevenly distributed in zygotic cytoplasm within the time between the injection of human mitochondria and the first cleavage; secondly, that the number of mtDNA segregation units in the entire amount of mitochondria injected (ca. $5 \cdot 10^2$) is about ten. Analysis of human mtDNA distribution among the organs of transmitochondrial animals showed that skeletal muscles, liver and gonads are the targets frequently hit.

In: Genetics Research Summaries
Editors: L. E. Hewitt and J. W. O'Neil
ISBN: 978-1-62257-646-3

Chapter 128

Trace DNA Analysis

Kaye N. Ballantyne*
Department of Forensic Molecular Biology,
Erasmus University Medical Center, Rotterdam, The Netherlands

Research Summary

Trace DNA has become a large part of the average forensic laboratories' workload. Remarkably low DNA amounts (<100pg) have been successfully analysed to obtain profiles from a wide range of sample types. Touched objects constitute the most common source of trace DNA, but any type of biological material present in low amounts may be considered as trace, including minute blood deposits, saliva residue on partially consumed food, or even epithelial cells from the interior surface of condoms. In addition to supplementing existing analysis techniques in serious crime cases, trace DNA can allow investigation of volume crime cases such as burglary or vehicle theft, where previously DNA evidence was not considered usable. However, despite the widespread use of trace DNA, at present there are very few specific validated methods. This has lead to controversy in the use of trace DNA, and particularly the low copy number amplification technique. It has been established that the use of existing methodology (developed for high-copy number samples) leads to significant levels of artefacts with trace DNA, including allele drop-out and drop-in, stutter, and allelic/locus imbalance. To minimize these, there are numerous modifications that can be made to existed methods to increase the success of trace DNA analysis. These include reduced extraction volumes, increased cycle number, reduced PCR volume, and increased injection time for capillary electrophoresis. In addition, new research shows that the introduction of techniques such as whole genome amplification, molecular crowding, and post-PCR purification can significantly increase success rates with trace DNA. It is clear that each step of the analysis procedure (including collection, extraction, amplification and fragment detection) can, and should, be optimized with regards to trace DNA. However, the use of increasingly sensitive trace DNA analysis techniques must bring an increasing

* Corresponding author: E-mail: k.ballantyne@erasmusmc.nl

awareness of the potential for contamination, both within the laboratory and at crime scenes. In particular, this has implications for the analysis of trace DNA from cold cases, which were not collected or stored with highly sensitive DNA detection techniques in mind. Although it is certain that trace DNA will continue to be used within forensic biology, there may need to be modifications to a wide range of practices to ensure accuracy and reliability.

In: Genetics Research Summaries
Editors: L. E. Hewitt and J. W. O'Neil
ISBN: 978-1-62257-646-3

Chapter 129

The Use of Mitochondrial DNA in Forensic Identification

C. D. Matheson
Lakehead University, Thunder Bay, Canada

Research Summary

Mitochondrial DNA has been used for the purposes of forensic identification since the mid 90s. Since then dedicated laboratories have been set-up and guidelines for the analysis and interpretation of mitochondrial DNA have been established. The most common method of analysis has been to amplify and sequence the mitochondrial DNA to generate a mitochondrial profile. This method has been applied to species identification used on forensically important animal species, and human identification whereby the mitochondrial DNA is used for forensic identification of low copy number DNA and degraded remains. The use of mitochondrial DNA does not allow discrimination to an individual like nuclear DNA as it can only identify maternally related individuals. However this has the potential for the exclusion of an individual in forensic identification.

In: Genetics Research Summaries
Editors: L. E. Hewitt and J. W. O'Neil

ISBN: 978-1-62257-646-3

Chapter 130

DNA-Wrapped Carbon Nanotubes in Aqueous Solution and Air

Masaru Tachibana*
Department of Nanosystem Science, Yokohama City University,
Yokohama, Japan

Research Summary

The production of DNA-wrapped single-wall carbon nanotubes (SWNTs) in aqueous solution has led to a bid advance for separation or isolation of SWNTs. The understanding of the effect of DNA-wrapping on physical properties of SWNTs is important for basic understanding of DNA-SWNT interaction as well as engineering applications of DNA-wrapped SWNTs. Raman scattering of SWNTs are significantly influenced by DNA wrapping. The Raman spectra of DNA-wrapped SWNTs in air strongly depend on relative humidity. In addition, the DNA-wrapping mechanism for smaller diameter SWNTs is different from that for larger ones. Such properties of DNA-wrapped SWNTs can provide further applications in air. In this chapter, first, the production of DNA-wrapped SWNTs is described. Next, the Raman spectral features of DNA-wrapped SWNTs in aqueous solution and air are reviewed. Finally, the morphologies of DNA-wrapped SWNTs are discussed.

* Corresponding author: E-mail: tachiban@yokohama-cu.ac.jp

In: Genetics Research Summaries
Editors: L. E. Hewitt and J. W. O'Neil
ISBN: 978-1-62257-646-3

Chapter 131

Analysis of Mitochondrial DNA Sequence Variations in Type 2 Diabetes Patients Characterized for Haplogroup Affiliation

Antonella Cormio[1], Francesco Milella[1], Maurizio Marra[2], Maria Pala[3], Angela Maria Serena Lezza[1], Anna Rita Bonfigli[2], Claudio Franceschi[4,5], Palmiro Cantatore[1,6] and Maria Nicola Gadaleta[1,6*]

[1]Dipartimento di Biochimica e Biologia Molecolare "E. Quagliariello", Università di Bari, Bari, Italy
[2]Unità di Diabetologia, INRCA Ancona, Italy
[3]Dipartimento di Genetica e Microbiologia, Università di Pavia, Pavia, Italy
[4]Centro Interdipartimentale "L. Galvani" (CIG), Università di Bologna, Bologna, Italy
[5]Centro di Biologia Molecolare e Genetica, INRCA, Ancona, Italy
[6]Istituto di Biomembrane e Bioenergetica CNR, Bari, Italy

Research Summary

The knowledge of the haplogroups of the analyzed subjects allows to prevent errors in the analysis of the association between mtDNA variations and some pathologies like diabetes. In this paper a cohort of type 2 diabetic Italian patients with known mitochondrial genetic background (haplogroup) were investigated for the presence in the blood of the 3243A>G mutation and other variants of close to $tRNA^{Leu(UUR)}$ gene (16S/ND1 region) of mitochondrial DNA (mtDNA). The 3243A>G mutation was present in 0.4% of analyzed patients in agreement with the reported frequencies of such mutation in other European groups. This

* Corresponding author: Tel. +39-080-5443309, Fax: +39-080-5443403, E.mail: m.n.gadaleta@biologia.uniba.it

result suggests a limited role for the 3243A>G mutation in the pathogenesis of diabetes also in the Italian population. In the 16S/ND1 region 15 base changes were found. The knowledge of the haplogroup of the analyzed individuals allowed to distinguish between haplogroup-specific polymorphisms and variants potentially associated to diabetes.

In: Genetics Research Summaries
Editors: L. E. Hewitt and J. W. O'Neil
ISBN: 978-1-62257-646-3

Chapter 132

Regulational Role of Mitochondria on Ageing and Senescence

Hiroshi Kondoh*[*1], *Haruhisa Inoue*[2], *Akifumi Takaori-Kondo*[3], *Takumi Mikawa*[1] and *Manabu Minami*[4]

[1]Department of Geriatric Medicine Graduate School of Medicine, Kyoto University, Shogoin, Sakyo-ku, Kyoto, Japan
[2]Center for iPS Cell Research and Application (CiRA)
Institute for integrated Cell-Material Sciences (iCeMS) Kyoto University, JST YAMANAKA iPS Cell Project Office, Shimogyo-ku, Kyoto, Japan
[3]Department of Hematology & Oncology Graduate School of Medicine, Kyoto University
[4]Department of Clinical Innovative Medicine Translational Research Center, Kyoto University

Research Summary

Mitochondria is the major supplier of intracellular ROS (Radical oxygen species). Since the proposal of free radical theory by Harman in 1956, there has been a long dispute about the impact of mitochondria on longevity. Recently it was reported that several mouse models of mitochondria mutant display the phenotype of premature ageing or age-related disease. Moreover ectopic expression of mitochondria-targeted radical scavenger extends organismal life span in mice.

These suggest that mitochondria dysfunction intriguingly correlates with ageing and senescence. However, opposed or different ideas on the role of mitochondria on longevity are also proposed in the research of yeast genetics. Here we review the recent advance on molecular link between mitochondria and senescence. We propose that high efficient

[*] Corresponding author: E-mail: hkondoh@kuhp.kyoto-u.ac.jp

mitochondria model might explain clearly the importance of mitochondria on ageing and senescence.

In: Genetics Research Summaries
Editors: L. E. Hewitt and J. W. O'Neil
ISBN: 978-1-62257-646-3

Chapter 133

Phylogeography and Speciation Processes in Marine Fishes and Fishes from Large Freshwater Lakes

***Michael Matschiner*[1]*, Reinhold Hanel*[2*] *and Walter Salzburger*[1]**
[1]Zoological Institute, University of Basel, Basel, Switzerland
[2]Institute of Fisheries Ecology, Johann Heinrich von Thünen-Institute, Hamburg, Germany

Research Summary

Fishes constitute about half of all known vertebrate species and have colonized nearly all available marine and freshwater habitats. The greatest diversity of fishes is found in the marine realm as well as in large (and often old) freshwater lakes such as the East African Great Lakes. Here, we compare the phylogeographic history of fishes in marine and large freshwater ecosystems, with particular emphasis on groups that underwent adaptive radiation, *i.e.* the emergence of a multitude of species from a single ancestor as a consequence of the adaptation to different ecological niches. Phylogeographic analyses are highly suited to identify and compare causal agents of speciation in rapidly diversifying groups. This is particularly true for fishes, in which distribution ranges and preferred habitat structures can be quantified in a straightforward matter.

* Author of correspondence.Reinhold Hanel E-mail: reinhold.hanel@vti.bund.de

In: Genetics Research Summaries
Editors: L. E. Hewitt and J. W. O'Neil

ISBN: 978-1-62257-646-3

Chapter 134

Phylogeography: Its Importance in Insect Pest Control

M. D. Ochando, A. Reyes, D. Segura and C. Callejas
Departamento de Genética, Facultad de CC. Biológicas,
Universidad Complutense, Madrid, Spain

Research Summary

Phylogeography involves knowledge of the spatial distribution of related individuals and historical information on the relationship within and among populations and species. The phylogeography of many groups has been studied over recent decades, and this field of knowledge is now becoming important in solving the problems of pest control in agriculture and forestry. An understanding of the nature of the genetic variation within and between pest populations is of paramount importance in the design of pest control programmes and their success – and successful programs are certainly needed since increasing levels of trade and passenger travel, the growth of new plant species in new regions and climate change are all assisting the spread of economically important insect pests.

Molecular genetics is now providing us with new and more sensitive tools for developing appropriate insect control and eradication strategies. Our group is studying the genetic variation of tephritids and whiteflies, both of which are important pests of agricultural and ornamental plants in the Mediterranean region (ecosystems). The information presented here is a summary of some of the results obtained in enzyme electrophoresis (MLEE), abundant soluble protein content, random amplification of polymorphic DNA (RAPD-PCR), intermicrosatellite (ISSR), restriction fragment length polymorphism (RFLP) and mtDNA sequencing analyses. These data document the phylogeographic structure of these taxa's populations, and provide information regarding their bioinvasion and colonisation of new areas, their geographic variation, and gene flow among regions. Such information could be helpful in the taking of pest management decisions, and highlights the need for the coordination of control programs between regional, national and international authorities.

In: Genetics Research Summaries
Editors: L. E. Hewitt and J. W. O'Neil
ISBN: 978-1-62257-646-3

Chapter 135

Host Specificity and Speciation in Parasitic Plants

Chris J. Thorogood and Simon J. Hiscock
University of Bristol, Bristol, UK

Research Summary

Parasitic plants have aroused curiosity among scientists for centuries, yet they remain one of the most poorly understood groups of flowering plants (angiosperms), and much of their evolutionary biology remains a mystery. There are approximately 4,000 species of parasitic plants (in 19 families) which occur in all major biomes, from arctic islands to tropical forests. The evolutionary shift to parasitism has been associated with the degeneration of morphological features traditionally used in plant classification to infer evolutionary relationships, making systematic studies of the relationships between parasitic plants and their photosynthetic ancestors extremely difficult. Therefore until very recently, our understanding of the evolutionary origins of parasitic angiosperms has lagged behind that of other major groups of angiosperms.

Parasitic plants show considerable variation in their host specificity. For example, some species infect hundreds of species from taxonomically diverse families, while others are restricted to a single host species. Host ecology may isolate parasite populations and facilitate genetic divergence. For example in leaf-eating (phytophagus) insects which, like parasitic plants, are discriminate users with respect to the host plants on which they feed, host specificity has been demonstrated to drive genetic divergence, and ultimately speciation. However speciation in parasitic plants has, until recently, remained relatively unexplored by evolutionary biologists, and our understanding of host specificity as a potential catalyst for speciation in these plants has lagged behind that of phytophagus insects. Now, in light of recent research, a similar pattern of host-driven speciation in parasitic plants is emerging. This chapter reviews recent research into the host specificity of parasitic plants, which appears to be an important and underestimated promoter of genetic divergence and speciation.

In: Genetics Research Summaries
Editors: L. E. Hewitt and J. W. O'Neil

ISBN: 978-1-62257-646-3

Chapter 136

Gene Flow, Genetic Drift, and Geographic Variation of the Ainu: An Assessment Based on Nonmetric Cranial Traits

Tsunehiko Hanihara *
Department of Anatomy and Biological Anthropology,
Saga Medical School, Saga, Japan

Research Summary

Gene flow and genetic drift are important factors affecting geographic variation of human phenotypic traits. In the present study, the effects of gene flow from an outside source on the pattern of within- and among-group variation of Hokkaido Ainu, one of the most generalized eastern Asian populations, are examined by applying R-matrix method to 24 nonmetric cranial traits. The R-matrix method is developed initially for single-locus traits and modified for use with quantitative (metric) phenotypic data. In this study, tetrachoric correlation coefficients estimated by maximum likelihood method, the threshold value for each trait estimated by univariate probit analysis, and the census population sizes of the regional groups of the recent Ainu were used for applying the R-matrix method to nonmetric morphological data. Within-group variance based on nonmetric data was estimated by bootstrap sampling method. The results obtained suggest the possibility of admixture between the immigrants from Northeast Asia as represented by the Okhotsk culture people and the indigenous inhabitants in Hokkaido during the 5th – 12th centuries A.D., at least in the coastal region along the Sea of Okhotsk. Such gene flow from Northeast Asian continent may have a certain degree of effect on genetic structure of recent Ainu as suggested from morphological and

* Correspondence to: Tsunehiko Hanihara, Department of Anatomy and Biological Anthropology, Saga Medical School, Nabeshima, Saga 849-8501, Japan. E-mail: hanihara@cc.saga-u.ac.jp

ancient mitochondrial DNA evidence. The present findings suggest, moreover, a possible inter-population relationship between the Ainu and non-Ainu Japanese at recent period of time. The present analyses provide results that can be interpreted in terms of archaeologically and historically suggested pattern of gene flow and isolation.

In: Genetics Research Summaries
Editors: L. E. Hewitt and J. W. O'Neil
ISBN: 978-1-62257-646-3

Chapter 137

Suture Zones and Phylogeographic Concordance: Are They the Same and How Should We Test for Their Existence?

Nathan G. Swenson
Center for Tropical Forest Science – Asia Program
Arnold Arboretum, Harvard University,
Cambridge, MS, US

Research Summary

Phylogeography uses present day geographic patterns of genotypes to infer the historical distribution and demography of species. While species-specific patterns have been interesting, repeated geographic patterns across species, known as phylogeographic concordance, provide evidence for general mechanisms and biogeographic events that have together shaped the distribution and diversity of genotypes. As the evidence for phylogeographic concordance started to accumulate, phylogeographers began to revisit the concept of suture zones. Suture zones, as originally described by Charles Remington, are geographic regions where multiple sister species pairs experience secondary contact and hybridize. Of interest was that some of Remington's suture zones seemed to include regions of phylogeographic concordance, suggesting that the concept and mechanisms behind the formation of suture zones could be translated from hybrid zone research to phylogeographic research. Thus, the challenge now is to explore whether a full integration of the concepts of suture zone formation and phylogeographic concordance is possible and how we should test for their existence. Here I will discuss the suture zone concept and how it relates to phylogeographic concordance, discuss previous attempts to statistically test for the existence of suture zones and phylogeographic concordance, and propose a more rigorous statistical analytical approach towards testing for the existence of suture zones and phylogeographic concordance.

In: Genetics Research Summaries
Editors: L. E. Hewitt and J. W. O'Neil

ISBN: 978-1-62257-646-3

Chapter 138

Differentiation History of Dragonflies in the Insular East Asia Revealed by the Gene Genealogy *(Odonata: Hexapoda)*

Takuya Kiyoshi*
Center for Ecological Research, Kyoto University
Hirano, Otsu, Shiga, Japan

Research Summary

The insular East Asia contains many islands encompasses various climatic regions, from subtropical to subarctic. Its fauna contains about 400 species of the order Odonata (dragonflies and damselflies), including many endemic ones. Two groups of dragonflies, the genus *Davidius* and *Anotogaster sieboldii* show interesting radiation patterns in this region. Both the mitochondrial and nuclear gene genealogy revealed that four species of the genus *Davidius* in this region seem to have diversified through the geographical connection and disconnection between the Korean Peninsula and Japanese main islands. The habitat of the larvae of *D. moiwanus* sspp. were estimated to have shifted from rivers to narrow streams in wetlands. Molecular phylogeographical analyses revealed not only their divergence history in the insular region but also including the process divided them from the continental congeners. Mitochondrial gene genealogy based on COI gene sequence data revealed that *A. sieboldii* includes two deeply differentiated clades that seem to have diverged in late Miocene or early Plicocene. Each of these two clades includes three inner clades that seem to have differentiated in Pleistocene.

* Telephone: (+77)-549-8215; Fax: (+77)-549-8201; E-mail: kiyoshi@ecology.kyoto-u.ac.jp

In: Genetics Research Summaries
Editors: L. E. Hewitt and J. W. O'Neil
ISBN: 978-1-62257-646-3

Chapter 139

Statistical Phylogeography, Ecological Niche Models and Predicting Glacial Refugia: An Examination of Key Assumptions

***Nathan G. Swenson*[1,*] *and Jason Pither*[2]**

[1]Center for Tropical Forest Science – Asia Program
Arnold Arboretum Harvard University, Cambridge, MS, US
[2]Biology and Physical Geography University of British Columbia Okanagan
Kelowna, British Columbia, Canada

Research Summary

Recent statistical phylogeographic work has sought to integrate ecological niche models (ENMs) into phylogeography as a methodology for testing and formulating hypotheses regarding the geographic histories of species. The lure of ENMs lies in their ability to use present day climatic and physiographic affinities of species to predict the past geographic distributions of these same species under different climatic regimes. These predicted distributions can then be used to evaluate the likelihood and location of glacial refugia. Given the bourgeoning interest in ENMs, now is an appropriate time for phylogeographers to reexamine their conceptual and methodological foundations. We argue that in the rush to assimilate this methodology, critical underlying assumptions have not been adequately considered. Here we discuss some of the fundamental assumptions underlying ENMs and how their violation may lead to faulty phylogeographic inferences. We will conclude by discussing alternative avenues for future research that will enable a more reliable and fruitful integration of ENM and phylogeography.

* Email: nswenson@oeb.harvard.edu

In: Genetics Research Summaries
Editors: L. E. Hewitt and J. W. O'Neil

ISBN: 978-1-62257-646-3

Chapter 140

The Need for a Multispecies, Multilocus Phylogeographical Approach

Cristiano Vernesi*

Centro di Ecologia Alpina – Research and Innovation Centre,
Fondazione Edmund Mach
Viote del Monte Bondone, Trento, Italy

Research Summary

Phylogeography focuses on the study of the geographical structure of gene lineages within single species. The aim to analyze how phylogenetic relationships of genealogical lineages are distributed across the geographical landscape makes phylogeography a discipline embedded in the wider field of biogeography which is primarily concerned with inference and, possibly, identification of the major processes shaping organismal diversity at different levels of geographical and taxonomical scales. The adoption of molecular data to infer patterns of genetic variation interpreted in a geographical context positively affected biogeographical research: many fruitful insights can be gained about the different roles of bottlenecks, population expansions, vicariance and gene flow in structuring genetic variability. From its inception, at least two major breakthroughs occurred in phylogeography: the advent of comparative phylogeography and the use of model-based statistical analysis which led to statistical phylogeography. The comparative approach involves the comparison among co-distributed taxa of geographical patterns of genetic variation thus allowing detailed inferences to be drawn about, for instance, landscape evolution, dispersal across a region, speciation, extinction and adaptive radiation. Linking population processes to regional biogeographical and diversity patterns can have important outcomes also for ecology and evolution studies. Historically and evolutionary independent regions can be identified, statistically testing their independence through adequate state-of-art analytical methods.

* E-mail: vernesi@cealp.it